W. J. Dobbs

Elementary Geometrical Statics

An Introduction to Graphic Statics

W. J. Dobbs

Elementary Geometrical Statics
An Introduction to Graphic Statics

ISBN/EAN: 9783337277413

Printed in Europe, USA, Canada, Australia, Japan

Cover: Foto ©berggeist007 / pixelio.de

More available books at **www.hansebooks.com**

ELEMENTARY
GEOMETRICAL STATICS

AN INTRODUCTION TO

GRAPHIC STATICS

BY

W. J. DOBBS, M.A.

SOMETIME FOUNDATION SCHOLAR OF ST. JOHN'S COLLEGE, CAMBRIDGE

𝕷onbon

MACMILLAN AND CO., Limited

NEW YORK : THE MACMILLAN COMPANY

1897

PREFACE.

It has long been my firm conviction that the teaching of *Elementary Statics* would gain in clearness and educational value by a more general use of geometrical methods. The fundamental propositions of the subject are essentially geometrical, but it is usual for the beginner to abandon the direct use of the geometrical methods in favour of the analytical formulae to which they give rise. This is a pity. Mathematical formulae fail to appeal to the eye with the direct force of a geometrical figure, and the power and neatness of the geometrical methods are unquestionable. The practical engineer makes considerable use of *Graphic Statics*, but the subject has been much neglected in this country, and there seems to be no book which leads up by easy stages to the mastery of a subject at once interesting and instructive, and which can be systematically dealt with in a scientific manner. In most recently published text-books on *Elementary Statics*, an attempt is made to deal with the subject of *Graphic Statics* in a short chapter or a few articles, but the matter is worthy of better treatment, and there seems to be a growing need for some such volume as the present, which deals with *Geometrical Statics* alone. The book is essentially an elementary one, and is intended to prepare the way for such works as Major Clarke's *Graphic Statics* or Professor Hoskins' *Elements of Graphic Statics.*

Rather against the advice of friends, I have not attempted to write a treatise independent of existing text-books. My wish is to supplement, not to compete with, such. Hence the *Principle of Transmissibility of Force* and the *Parallelogram of Forces* have been assumed, and a direct plunge taken into the geometrical aspect of the subject. For the groundwork and, later on, for an exposition of the *Laws of Friction*, the student is referred, by permission, to Professor Loney's *Elements of Statics.*

In preparing this work, I have consulted most available English books which bear upon the subject, and, in particular, Professor Hoskins' *Elements of Graphic Statics.* The method of lettering the diagrams is the extension of Bow's notation adopted in that volume.

Each chapter concludes with a number of worked-out examples, which are followed by a set of exercises for the student. Each set contains a collection of numerical examples, followed, in most cases, by others of a more general kind, which are intended to be worked with the aid of elementary pure geometry. The numerical examples are designed primarily for solution by means of accurately drawn figures; a careful worker, however, can in the more simple cases obtain fairly accurate results by freehand drawing, while the student of Trigonometry can calculate the lengths of the lines of his force diagram, and thus obtain accurate solutions.

I may claim most of the examples as my own original problems, accumulated during the last six years while teaching the subject to Woolwich pupils. Those which are not original are taken, for the most part, from recent examination papers set to candidates for admission to the Royal Military Academy.

The figures have, in most cases, been reduced in size from my original drawings, so as to admit of *space diagram* and *force diagram* being placed side by side on the same page. Those, however, which constitute the answers to numerical questions, are reproduced, in general, on the scale in which they were originally drawn. This has, in some cases, necessitated corresponding figures being placed on different pages facing each other. Attention is drawn to the numbering of the figures. Corresponding to the *space diagram* 108, we have the *force diagram* 108*a*, etc.

In conclusion, I take this opportunity of tendering my warmest thanks to several mathematical friends, to whom I am indebted for much kindly encouragement and assistance. My former mathematical master, the Rev. Henry Williams, read through the work in manuscript, and again as it went through the press, and

to his personal interest I owe much. I desire also, in
particular, to acknowledge my obligations to my
friend and former colleague, Mr. R. W. Bayliss, late
scholar of St. Peter's College, Cambridge, who has
also read through the whole of the proof sheets, and
to whom I am indebted for many criticisms and
suggestions.

Any corrections or suggestions for improvement
from either teachers or students will be most thank-
fully received. I have spared no pains in working out
the answers to the examples, and hope that no serious
errors will be found to have escaped correction.

W. J. DOBBS.

3 Sunningdale Gardens,
    Kensington.
        July, 1897.

CONTENTS.

CHAPTER I.

FUNDAMENTAL PRINCIPLES.

1. Geometrical Representation of Force.

A force is completely determined when we know (i.) its point of application, (ii.) the direction in which it acts, and (iii.) its magnitude.

Now the point of application has a corresponding point in the diagram which represents the material system under consideration; and from that point a straight line may be drawn in the direction which represents, in the diagram, the direction in which the

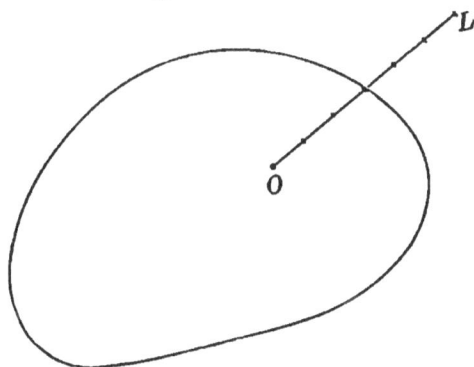

Fɪɢ. 1.

force acts; further, this line may be drawn of such a length as to represent the magnitude of the force, by making it contain, on any suitable scale, as many units of length as the force contains units of force.

A

Thus, in the accompanying figure, we have a diagram representing some material system, and the point O of the diagram represents a material point of the system. The straight line OL is drawn in such a direction that it represents, relatively to the rest of the system, the direction of a force whose measure is 6 applied at the point O; and OL is taken 6 units of length, and in this way represents graphically the magnitude also of the force.

2. Now it is to be noticed that, in the above figure, we have two different scales. The outline of the figure and the position of the point O represent the configuration of the material system under consideration, on a scale in which length represents length—for instance, one inch may be taken to represent one foot; while the line OL does *not* represent a material line of the system, for in this part of the diagram length represents force—for instance, one inch may be taken to represent the weight of one pound.

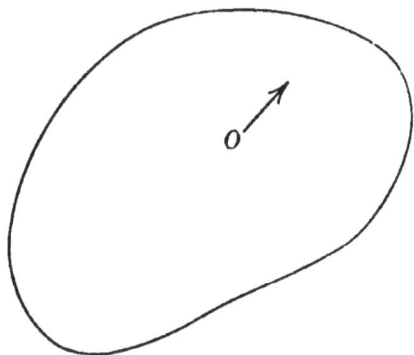

FIG. 2.

FIG. 2 a.

In order to avoid the confusion which would otherwise arise when a number of forces are represented in this way, it is found convenient to draw two separate figures—one, a diagram of the material system under consideration, in which length represents length, called a *space diagram*; and the other, a diagram in which lines represent forces, called a *force diagram*.

Thus HL, 6 units long, represents in the force diagram a force whose measure is 6 applied at the point which is represented by $O˙$ in the space diagram.

3. Principle of the Transmissibility of Force.

We take it as axiomatic, that two *equal* forces acting in opposite directions at two points A, B of a rigid body, so that the force acting at A is in direction AB, and that acting at B in direction BA, produce no effect upon the body as a whole. The tendency is merely to compress the portions of the body between A and B, and as we are dealing with an ideally rigid body, that is, a body in which the several parts are so inseparably connected that they retain the same positions with regard to one another under all circumstances, the effect upon the body as one solid piece is *nil*. Similarly, if the two equal and opposite forces act outwards instead of inwards, they produce no effect upon the body as a whole. From this axiom we deduce, as in Art. 19 of Loney's *Elements of Statics*, the Principle of Transmissibility of Force, which states that a force acting at A in the direction AB has the same effect upon the body as a whole as an equal force acting in the same direction at any point of the rigid body situated in AB or AB produced either way. Thus there are a succession of points of the body, all situated in the same straight line, any one of which may be considered to be the point of application of the force. This line is called the *line of action* of the force, and the force is said to *act along* its line of action. Further, the line of action may be extended beyond the limits of the body, and the force considered as applied at a point outside the actual body altogether, if we suppose the body to be ideally extended so as to include this point, which must be treated as a point of the body.

4. In practice, in representing a force geometrically, we do not trouble about indicating the point of application. In the space diagram we draw a line XY showing the line of action of the force, and insert an arrow to indicate the direction in which the force acts along its line of action; and in the force diagram another line HL, parallel to XY, shows graphically the magnitude of the force. The force is described as acting along XY when its direction is from X towards Y, and as acting along YX when its direction is from Y towards X.

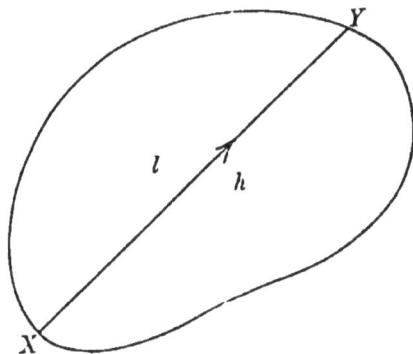
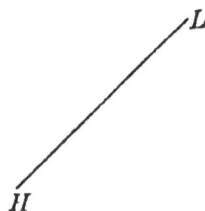

FIG. 3.   FIG. 3 a.

A very convenient notation, and one which will afterwards be found to be extremely useful, is indicated in the figure. The letter h is placed on one side of the line XY, and the letter l on the other; then the straight line separating the spaces marked h and l respectively is called the line hl, and the force which acts along the line hl is represented in the force diagram by HL.

5. The converse of the axiom above referred to (Art. 3) is equally important; namely, two forces acting upon a rigid body cannot balance one another unless they are equal and opposite and act in the same straight line. This we also take as axiomatic, and it leads to the converse of the principle of transmissibility of force, which is as important as the principle itself. It is this, —if a force acting at A has the same effect upon a rigid body as a whole as another force acting at B, then B must be a point in the line of action of the first force, and the two forces must be equal and in the same direction.

6. The Parallelogram of Forces.

If two forces act along, and are represented by, the two sides of a parallelogram drawn from one of its angular points, their resultant acts along, and is represented by, the diagonal of the parallelogram drawn from that angular point.

Thus, if a force whose measure is P acts along OA and is represented by OA, so that OA contains P units of length; and if a force whose

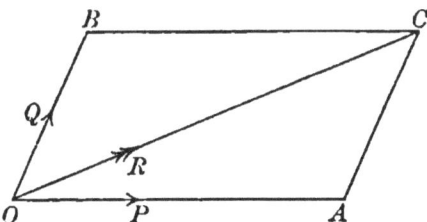

FIG. 4.

measure is Q acts along OB and is represented by OB, so that OB contains Q units of length; then, completing the parallelogram $OACB$, the resultant of P and Q acts along OC and is represented by OC.

Hence, if OC contains R units of length, the resultant of P and Q is a force whose measure is R acting along OC.

This gives the following method for finding graphically the resultant of two given forces: -

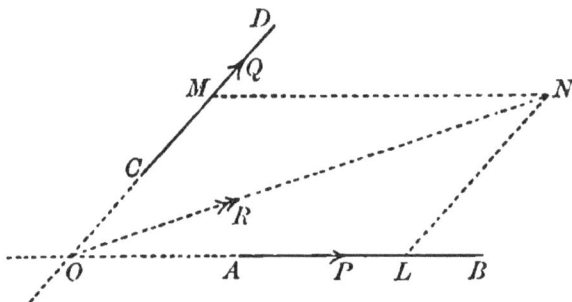

FIG. 5.

Let two given forces whose measures are P and Q act along the given lines AB, CD; it is required to find their resultant. Let the given lines AB, CD intersect at O. Then both forces may be supposed to act at O. Along OB measure OL to contain P units of length, and along OD measure OM to contain Q units of length. Complete the parallelogram $OLNM$ and measure ON. Suppose ON contains R units of length. Then the resultant of the two forces acts along ON and its measure is R. Its point of application may be taken to be any point in ON, or ON produced either way.

For the proof of this very important proposition see Loney's *Elements of Statics*, Art. 43. The student should notice that the two forces are represented by OL, OM, both drawn *away from* O, and that the resultant is intermediate in direction to the directions of P and Q.

The force whose measure is R, and whose line of action is ON, is not an actual force applied to the body; it is, rather, an ideal force which may be conceived to replace the given forces in their effect upon the body as a whole. We cannot locate the point of application of the resultant, although we are able to determine its line of action. This line of action may fall altogether outside the limits of the body on which the forces act. In such a case the interpretation is, that if the body be supposed to be ideally extended so as to include a portion of the line ON, the given forces may be conceived to be replaced by a force whose measure is R, applied, in the direction of ON, at a point of ON supposed to be rigidly connected with the body.

7. In the above direct use of the Parallelogram of Forces it will be noticed that we have our space diagram and our force diagram in one; in fact the force diagram has been constructed over the space diagram. This, in practice, would cause a great amount of confusion; but we can separate the two diagrams thus:

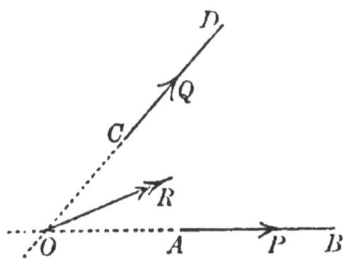
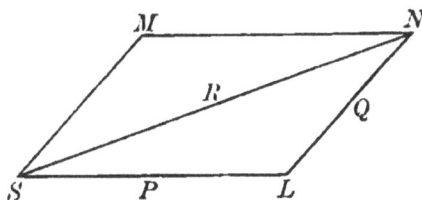

FIG. 6.   FIG. 6 a.

Instead of measuring a line along OB to represent the force P, draw SL in the direction of the force P to contain P units of length; also draw SM in the direction of the force Q to contain Q units of length.

Then, completing the parallelogram $SLNM$, the resultant is represented in magnitude and direction by SN, and its line of action is a straight line through O drawn parallel to SN.

8. But now we notice that we do not require to draw the whole parallelogram in the force diagram. All that is necessary is to draw one half of it, namely the triangle SLN.

Hence, finally, we have the following simplified method: '

To find graphically the resultant of two given forces.
Let P, Q be the measures of two given forces acting along the two given lines AB, CD respectively.

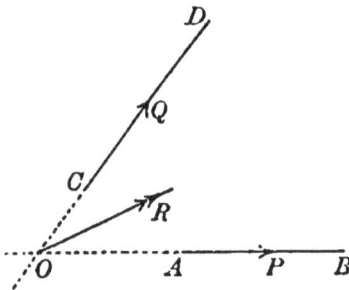 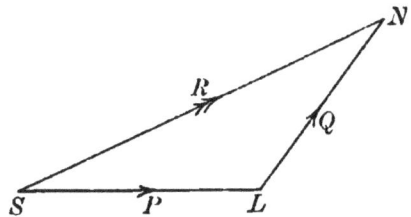

FIG. 7.   FIG. 7 *a*.

Starting from some suitable point S, and with any suitable scale, draw SL equal to P units of length in the direction of the force P. This takes us to the point L. From L draw LN equal to Q units of length in the direction of the force Q. Then the straight line from S, where we started, to N, where we finished, represents in magnitude and direction the resultant of the two given forces. We measure SN and find it is (say) R units of length. Find O, the point of intersection of AB and CD. Then the resultant is R units of force and acts along a line through O drawn parallel to *SN*.

This is the fundamental method of constructing the resultant of two given forces, and is the foundation of *Graphic Statics*. We see that the method fails when the point O is inaccessible. We will return to this case in a future chapter.

It is to be particularly noticed that the angle *SLN* is the *supplement* of the angle between the directions of the forces P, Q.

9. *Conversely, to resolve a given force into two components in two given directions.*

Let OL be the line of action of the given force whose measure is P. We may take any point O in

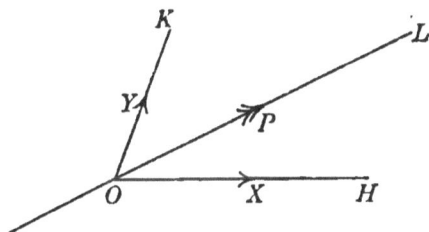
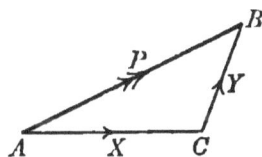

FIG. 8.   FIG. 8 a.

OL as its point of application. Let OH, OK be straight lines through O in the given directions.

Draw AB in the direction of OL and of length equal to P units; then draw AC parallel to OH, and BC parallel to KO, meeting in C.

Measure AC, CB. Let AC contain X units, and CB Y units; then X, Y are the measures of the components required along OH, OK respectively. For, by the preceding piece of work, the resultant of X and Y is represented by AB and acts along OL.

10. *If two forces acting along OA and OB are represented by m times OA and n times OB respectively, their resultant acts along OC, and is represented by m+n times OC, where C is a point in AB such that m times AC=n times CB. .*

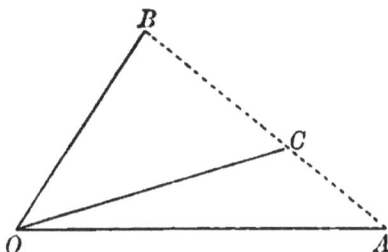

FIG. 9.

The force which acts along *OA* is equivalent to two forces *acting through* O represented by *m* times *OC* and *m* times *CA* respectively.

The force which acts along *OB* is equivalent to two forces *acting through* O represented by *n* times *OC* and *n* times *CB* respectively.

Let the two given forces be replaced by these two pairs of components. Then the two forces represented by *m* times *CA* and *n* times *CB*, *both acting at* O, balance one another, and can therefore be removed. Also the two forces represented by *m* times *OC* and *n* times *OC*, both acting along *OC*, are equivalent to a single force represented by *m+n* times *OC* acting along *OC*.

Hence the resultant acts along *OC* and is represented by *m+n* times *OC*.

In particular, if the forces are represented by *OA* and *OB*, their resultant acts along *OC* and is represented by twice *OC*, where *C* is the middle point of *AB*.

11. Ex. 1. *Find the resultant of two forces* 43 *and* 21 *pounds' weight acting at an angle of* 105½°.

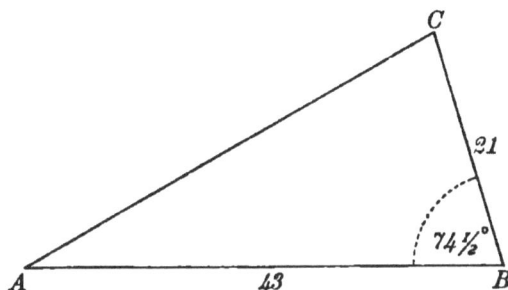

FIG. 10.

With any suitable scale make AB of length 43 units. Make angle ABC = supplement of 105½° = 74½°, and make BC of length 21 units. Join AC. Then, on measurement, AC is found to be of length 42·5 units, and the angle BAC of magnitude 28½°. Hence the resultant is 42·5 pounds' weight, making an angle of 28½° with the direction of the first force.

12. Ex. 2. *Two forces, one of which is of given magnitude, are inclined at a given angle. Show how to find the second force in order that the resultant may be of given magnitude.*

Taking AB to represent the given force to scale, make the angle ABX equal to the supplement of the given angle, BX being taken of unlimited length. Then the second force will be represented by some line BC taken along BX. To get the position of C we describe a circle with its centre at A, and its radius of such a length that it represents to scale the given magnitude of the resultant.

The points, if any, in which the circle intersects BX are possible positions of the point C. If, as in the figure, the circle cuts BX in two points C_1, C_2, then measuring BC_1, BC_2, we have two possible values for the force required.

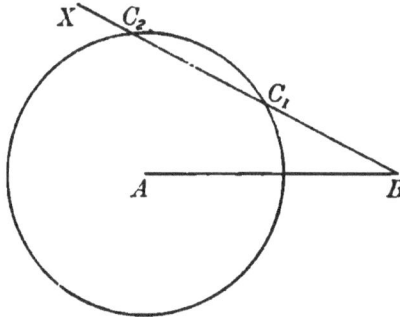

FIG. 11.

Thus, we see that the problem may be ambiguous, admitting of two solutions.

13. Ex. 3. *The resultant of two forces is equal to one of them. Show that, if this force be doubled, the new resultant is at right angles to the other force.*

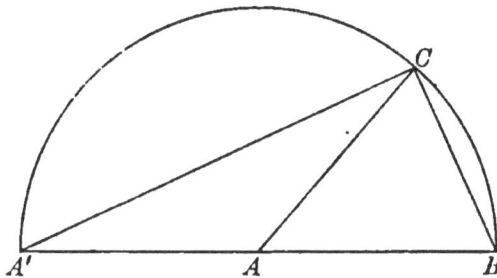

FIG. 12.

Let AB, BC represent the two forces. Then AC represents their resultant, and must, in the case before us, be equal to AB.

If the first of the two forces be doubled, the force, so altered, will be represented by $A'B$ where A is the middle point of $A'B$.

Thus A', C, B are points on a circle of centre A; ∴ angle $A'CB$, being an angle in a semi-circle, is a right angle, *i.e.* the new resultant is at right angles to the force represented by BC.

14. Ex. 4. *Show that the resultant of two forces $P+Q$ and P, acting at 120°, is of the same magnitude as the resultant of two forces $P+Q$ and Q, acting at the same angle.*

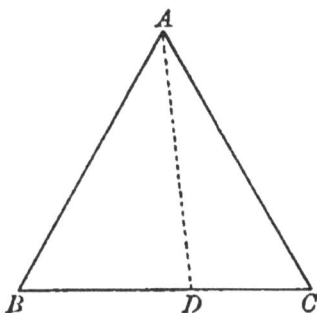

FIG. 13.

Take BDC a straight line, so that BD and DC are respectively P and Q units of length, and on it describe the equilateral triangle ABC.

Then AD represents in magnitude and direction

(i.) the resultant of forces represented by AB, BD; also (ii.) the resultant of forces represented by AC, CD. That is, AD represents

(i.) the resultant of forces $P+Q$ and P acting at 120°; also (ii.) the resultant of forces $P+Q$ and Q acting at 120°.

∴ the resultant of the first pair is equal in magnitude to the resultant of the second pair.

15. Ex. 5. *If one of two equal forces be reversed and doubled, the other remaining unaltered, it is*

found that the magnitude of the resultant is unaltered.
Find the original angle between the forces.

Let two equal forces P be represented by AB, BC, so that their resultant is represented by AC. Produce CB to D making $BD=2CB$. Then AD represents the resultant of one of the forces P unaltered, and the other doubled and reversed.

Hence, by the question,
$$AD=AC,$$
and therefore angle $ACB=$ angle ADB.

Bisect DB in E. Then
$$DE=EB=BC.$$

In the triangles ADE, ACB the sides AD, DE and the included

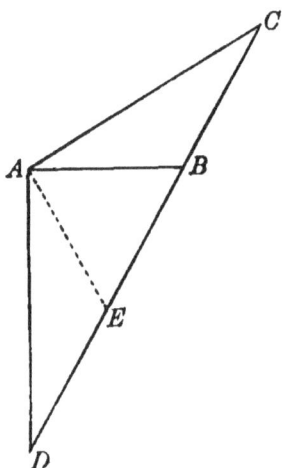
FIG. 14.

angle D are respectively equal to the sides AC, CB and the included angle C;
$$\therefore\ AE=AB;$$
$\therefore$ ABE is an equilateral triangle;
$\therefore$ the original angle between the forces is 60°.

16. Ex. 6. *Two given forces act in one plane at two given points of a rigid body; if they are turned round those points in the same direction through any two equal angles, show that their resultant will always pass through a fixed point.*

Let two forces, P and Q, act at two fixed points, H and K, in the directions OH and OK respectively, O being the point of intersection of their lines of action.

Take AB, BC to represent the forces P and Q respectively. Then AC represents their resultant R, which acts along a line through O drawn parallel to AC.

Now let the forces be turned in the same sense round H and K through the same angle, so that their new lines of action meet at O'. Then, since angle OHO'=angle OKO', the locus of O' is the circle described through H, K, O; and angle $HO'K$=angle HOK, so that the forces are inclined at the same angle as before.

If we suppose the triangle ABC altered so as to become the force triangle for the new position of the

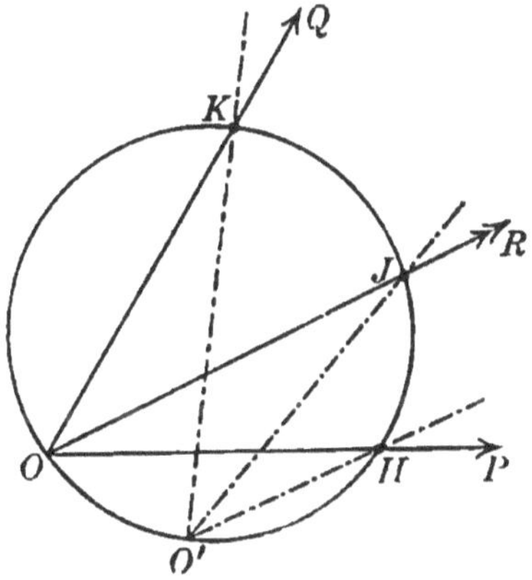

FIG. 15. 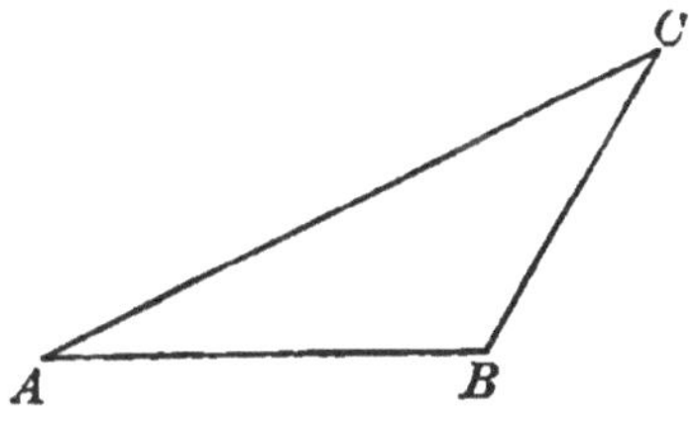   FIG. 15 a.

forces, the lines AB, BC remain of the same lengths as before, and contain the same angle. But two sides and the included angle are sufficient to determine the triangle in size and shape. Hence the resultant remains of the same magnitude as before, and is inclined to its components at the same angles as before. If, then, OJ is the line of action of the resultant in the first position, and J the point where this line meets the circle *HKO, O'J* must be the line of action of the resultant in the new position.

Thus we see that the resultant passes through the fixed point *J*, it remains of the same magnitude as before, and it has turned through the same angle as either of its components.

EXAMPLES I.

1. Find the resultant of

    (i.) 8 and 3 pounds' weight acting at an angle of 120° ;

    (ii.) 8 and 5 pounds' weight acting at an angle of 120° ;

    (iii.) 5 and 3 pounds' weight acting at an angle of 60° ;

and in each case give the angle that the resultant makes with the larger force.

2. *ABCDEF* is a regular hexagon. Find the magnitude, direction, and position of the resultant of forces of 4 pounds' weight acting along *FB*, and 2 pounds' weight acting along *AE*.

3. *ABC* is a triangle such that *AB*=3 inches, *BC*=4 inches, *CA*=5 inches. If 1 inch be taken to represent a force equal to the weight of 1 pound, find the magnitude, direction, and position of the resultant of two forces acting along and represented by *AC* and *CB* respectively.

Find also the magnitude, direction, and position of the resultant of 4 pounds' weight acting along *AC*, and 5 pounds' weight acting along *CB*.

4. *ABC* is a triangle, having its sides *BC, CA, AB* of lengths 14, 13, 15 inches respectively. Two forces, of magnitudes 25 and 39 pounds' weight, act along the lines *AB* and *AC* respectively. Find the magnitude, direction, and position of their resultant.

5. Find the resultant of forces of 200 and 100 pounds' weight acting at an angle of 60°.

6. Find the resultant of forces 15·8 and 23·7 pounds' weight acting at an angle of 113½°.

7. Resolve a force of 8 pounds' weight into two components, one of which is 3 pounds' weight in a direction making 60° with the given force.

8. Find the angle at which two forces of 16 and 20 pounds' weight must be inclined, in order that their resultant may be 33 pounds' weight.

9. Resolve a force of 31 pounds' weight into two components, making 98° and 40° with it on opposite sides.

10. The resultant of two forces P and Q is 8 pounds' weight, and makes an angle of 60° with the direction of P. If Q is 7 pounds' weight, determine P, and account for the double result.

11. The resultant of two forces, which act at an angle of 120°, is 31 pounds' weight, and one of the forces is 35 pounds' weight. Find the other force, and account for the double result.

12. Find the resultant of two equal forces acting at an angle of 120°.

13. If the magnitudes of two forces are given, their resultant is greatest when they act in the same direction, and least when they act in opposite directions.

14. The greatest and least resultants of two forces, of constant magnitudes, are given. Show how to find their resultant when they are inclined at a given angle.

15. E is a point in the side AB of the parallelogram $ABCD$. Show that the resultant of the two forces, represented in magnitude, direction, and position by CA and ED, is parallel to one of the sides of the parallelogram. Find also the line of action of the resultant.

16. If D is the middle point of the base BC of a triangle ABC, and the resultant of forces represented by BA, BD is equal to the resultant of those represented by CA, CD, show that the triangle ABC is isosceles.

17. It is required to apply to a given point two forces of given magnitudes, in order that their resultant may be of given magnitude and in a given direction. Explain how the directions of the two forces may be determined by geometrical construction. Under what circumstances does the construction fail?

18. A given force is to be resolved into two components, one of which is of given magnitude and acts in a given direction. Explain how the magnitude and direction of the other component may be determined by geometrical construction.

19. One of two forces is fully known, and the direction of the other is known. Show how to find the magnitude of the second, in order that the resultant may be in a given direction.

20. One of two forces is fully known, and the magnitude of the other is known. Show how to determine the direction of the second, in order that the resultant may be in a given direction.

21. Show how to resolve a given force into two others, one of which is of given magnitude, and the other in a given direction.

22. Show how to resolve a given force into two components, such that their sum may be of given magnitude and one of them in a given direction.

23. The resultant of two forces, one of which is fully known, is of given magnitude. If the known force be reversed, the resultant is of another given magnitude. Show how to determine the other force.

24. The resultant of two forces P and Q is in a direction perpendicular to that of P. Show that if P be doubled, Q remaining unaltered, the new resultant will be equal in magnitude to Q.

25. A straight line DE is drawn parallel to the base BC of a triangle ABC to meet the sides AB, AC in D, E respectively. Show that the resultant of forces represented by BE and DC is equal to a force represented by a line parallel to BC and equal to the sum of BC and DE.

26. Two forces P and Q act at an angle of 60°. Show that the magnitude of the resultant is unaltered if *either* of the given forces be replaced by a force $P + Q$ acting in the opposite direction.

27. The sides AB, BC, CD, DA of the quadrilateral $ABCD$ are bisected at E, F, G, H respectively. Prove that the resultant of the two forces acting along, and represented by, EG and HF is represented in magnitude and direction by AC. What is the line of action of the resultant?

28. Two opposite sides AB, CD of the quadrilateral $ABCD$ are bisected at E and F respectively. Prove that the resultant of forces acting along, and represented by, AC and BD is represented in magnitude and direction by twice EF. What is the line of action of the resultant?

29. Two forces are represented in magnitude and direction by OA and $2OB$. Show that their resultant is represented by $3OC$, where C is one of the points of trisection of AB.

30. OA, OB, OR represent in magnitude, direction, and position two forces and their resultant. If OC and OD be two equal lines cut off from OA and OB respectively, and if OR meet CD in G, find the ratio of CG to GD.

CHAPTER II.

FORCES ACTING AT A POINT.

17. *To find the resultant of any number of given forces acting at a point in one plane.*

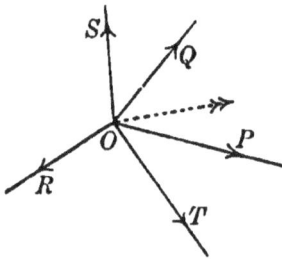

FIG. 16.

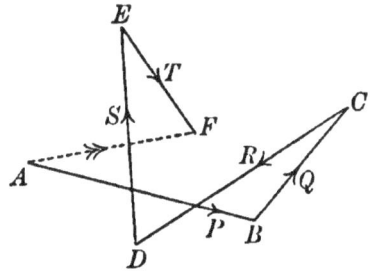

FIG. 16 a.

Let P, Q, R, S, T be the measures of five forces acting in known directions in one plane at the point O.

From any suitable point A, and with any suitable scale, draw AB in the direction of the force P and of length P units. This takes us to the point B. From B draw BC in the direction of the force Q and of length Q units. This takes us to the point C. Similarly draw CD, DE, EF in the directions of R, S, T respectively and of lengths R units, S units, T units respectively. Thus, finally, we arrive at the point F.

The straight line from A where we started, to F where we finished, represents the resultant in magnitude and direction, and its point of application is O.

For, the resultant of P and Q acts at O and is represented by AC. Let P and Q be replaced by their resultant. The resultant of this force and R acts at O and is represented by AD; therefore the resultant of P, Q, and R acts at O and is represented by AD.

Proceeding in this way, we see that the resultant of the whole system acts at O and is represented by AF.

The method is applicable to any number of forces, and the forces may be taken in any order. Also the lines in the force diagram may cross and recross one another any number of times. It is only necessary that the arrows in the force diagram should go one way round.

18. The student will, in the following manner, be able to satisfy himself that he gets the same result in whatever order he takes the forces. Let it be required to find the resultant of three forces, whose measures are P, Q, R, acting in known directions at the point O.

As before, take AB_1, B_1C_1, C_1D to represent P, Q, R respectively in magnitude and direction.

If we had taken the forces in the order P, R, Q, we should have obtained the figure AB_1C_2D, thus completing the parallelogram $B_1C_1DC_2$.

If we had taken the forces in the order Q, P, R, we should have obtained the figure AB_2C_1D, thus completing the parallelogram $AB_1C_1B_2$.

So Q, R, P gives AB_2C_3D, completing the parallelogram $B_2C_1DC_3$. R, P, Q gives AB_3C_2D, completing the parallelogram $AB_1C_2B_3$, and R, Q, P gives AB_3C_3D, B_3C_3 clearly being equal and parallel to AB_2.

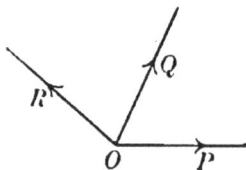
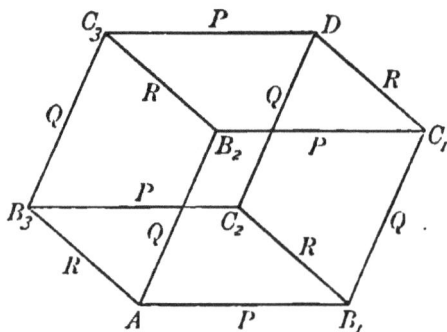

FIG. 17.   FIG, 17 a.

Thus we see that if we start at the point A, we always finish up at the point D, in whatever order we take the forces.

The resultant of the system acts at O and is represented by AD.

19. *Equilibrium of a system of forces acting at a point in one plane.*

If, in the force diagram of Art. 17, the point F coincides with the point A, the resultant of the system vanishes. In this case, replacing the forces P, Q, R, S by their resultant represented by AE acting at O, we see that the system is equivalent to two forces acting at O, the one represented by AE, and the other, T, represented by EA. Thus the system reduces to two equal and opposite forces acting in the same straight line. Therefore the forces are in equilibrium.

Hence we have the proposition known as

The Polygon of Forces.

If any number of forces, acting at a point, be represented in magnitude and direction by the sides of a polygon taken one way round, the forces are in equilibrium.

This of course includes, as a particular case,

The Triangle of Forces.

If three forces, acting at a point, be represented in magnitude and direction by the sides of a triangle taken one way round, the forces are in equilibrium.

It is to be particularly noticed that, in the Polygon of Forces, the polygon is essentially a *force diagram.* The forces do not *act along* the lines which represent them. So also in the Triangle of Forces.

20. *Conversely,* if a system of forces acting at a point in one plane be in equilibrium, and a force diagram be constructed, so that the forces are represented by straight lines each commencing where the preceding line ends, the arrows going one way round, then the last point must coincide with the first.

For, otherwise, the system would be equivalent to a resultant represented in magnitude and direction by the straight line drawn from the first point of the force diagram to the last point.

It is sometimes said that *the converse* of the Polygon of Forces is not true. But here we have *a true converse,* namely :

If a number of forces, acting at a point, be in equilibrium, it is possible to construct a closed polygon, whose sides taken one way round shall represent the forces in magnitude and direction.

This, of course, includes the following converse to the Triangle of Forces:

If three forces, acting at a point, be in equilibrium, it is possible to construct a triangle whose sides taken one way round shall represent the forces in magnitude and direction.

21. But there is a more general converse to the Triangle of Forces, namely:

If three forces, acting at a point, be in equilibrium, and straight lines be drawn parallel to their lines of action so as to form a triangle, then the sides of the triangle are proportional to the forces to which they are respectively parallel.

For, let P, Q, R be the measures of three forces in equilibrium, acting at the point O. Take BC in the direction of P and make it P units of length. Take CA in the direction of Q and make it Q units of length.

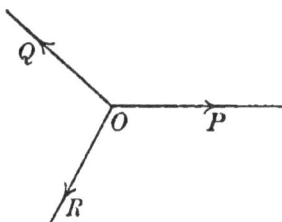

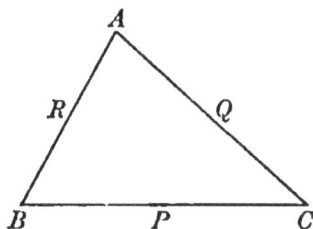

FIG. 18.   FIG. 18 a.

Then the straight line drawn from A in the direction of R and of length R units, must terminate at B; otherwise the forces would not be in equilibrium.

Now any triangle drawn with its sides parallel to the lines of action of P, Q, R will be similar to the triangle ABC, and will therefore have its sides proportional to P, Q, R.

In this wider sense the converse of the Polygon of Forces is not true, since two polygons with their sides respectively parallel are not necessarily similar.

22. *Three forces, of given magnitudes, act at a point in one plane. It is required to determine how these forces must be arranged so as, if possible, to produce equilibrium.*

Let P, Q, R be the given measures of the forces, and O the point at which they act.

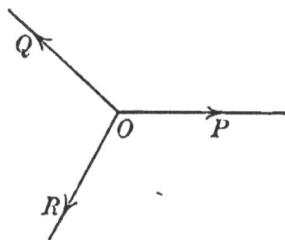
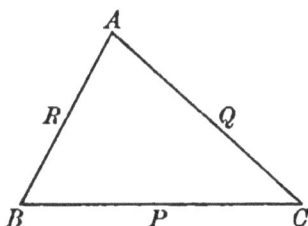

FIG. 19.   FIG. 19 a.

Construct a triangle ABC whose sides BC, CA, AB are of lengths P, Q, R units respectively.

From O draw straight lines in the directions of BC, CA, AB. Then, if the forces P, Q, R be arranged to act in these directions respectively, they will, by the Triangle of Forces, produce equilibrium.

This determines the relative directions of the three.

The method fails if any one of the forces is greater than the sum of the other two, as no triangle can be constructed having one side greater than the sum of the other two. In this case the problem is impossible of solution.

If one of the forces, P, be equal to the sum of the other two, the triangle becomes a straight line, the point A falling in BC. This shows that Q and R must be arranged to act in one and the same direction, and P in the opposite direction.

The student should notice that the angle between the directions of any two of the forces is the *supplement* of the corresponding angle of the triangle.

23. Let P, Q, R, S be the measures of known forces acting in known directions at a point O; and let X, Y be the measures of two other forces acting at O, at present unknown in magnitude or direction or both, which preserve equilibrium with the known forces; all the forces being in one plane.

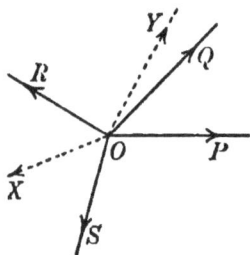

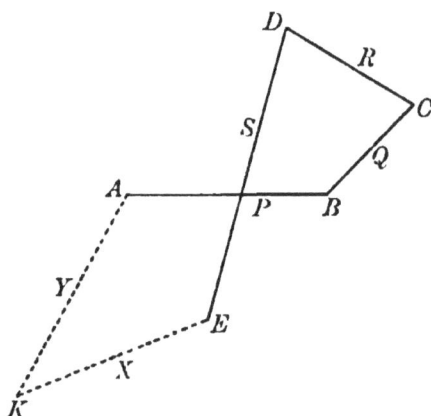

FIG. 20.   FIG. 20 a.

We can plan out the known forces in a force diagram at once. Thus, take AB, BC, CD, DE in the directions of P, Q, R, S and of lengths P, Q, R, S units respectively. This takes us from A to E.

In completing the force polygon, we shall have to go from E to A in two steps, as EK, KA, where EK represents X in magnitude and direction, and KA represents Y.

To complete the figure, we must know

either (i.) both magnitude and direction of one of the remaining two forces, say X;

or, (ii.) the directions of both of the remaining forces;

or, (iii.) the direction of one, say X, and the magnitude of the other, Y;

or, (iv.) the magnitudes of both X and Y.

For, (i.) suppose the force X is known completely. Then we can draw EK, and joining K to A we have a straight line which represents the remaining force Y in magnitude and direction.

Here, we see, we always get one solution, and one only.

(ii.) Suppose the directions of X and Y are both known, but not their magnitudes.

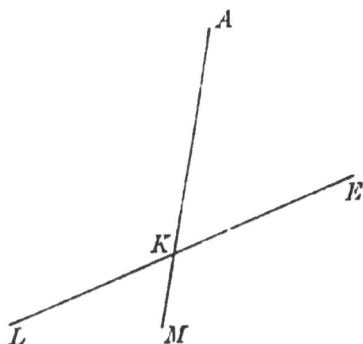

FIG. 21.

Draw EL in the direction of X, AM in the direction opposite to that of Y, and let EL, AM intersect at K. Then, measuring EK, KA, we have X, Y respectively.

It may be that LE has to be produced through E, in order to meet AM. In this case X is negative. Or, MA may have to be produced through A to meet EL. In this case Y is negative.

Here, also, we always get one solution, and one only.

(iii.) Suppose the direction of X is known and the magnitude of Y.

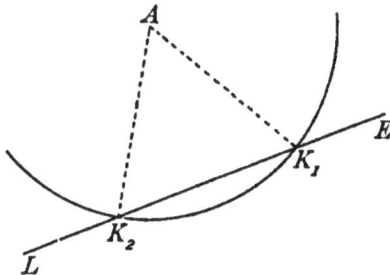

FIG. 22.

Draw EL in the direction of X, and with centre A, and radius whose measure is Y, describe a circle, which may cut EL in two points K_1 and K_2, giving two solutions.

Measuring EK_1 we have one value of X, and K_1A is the corresponding direction of Y. EK_2 gives another value of X, and K_2A is the corresponding direction of Y.

The circle may touch the line EL, in which case the two solutions coincide; or, the circle may not meet the line, in which case there is no solution.

(iv.) Suppose the magnitudes of X and Y are known. With centre E, and radius whose measure is X, describe a circle, and with centre A, and radius whose measure is Y, describe another circle.

These circles may or may not intersect in two points. Suppose they intersect in two points K_1, K_2. Then, again, we have two solutions.

The directions of X, Y are *either* those of EK_1, K_1A respectively, *or* those of EK_2, K_2A respectively.

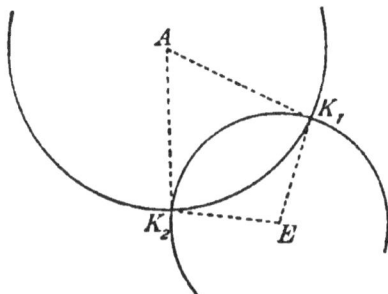

FIG. 23.

The two solutions coincide when $X = Y$, for then EK_1AK_2 is a parallelogram.

It appears, then, that if we have a number of forces in equilibrium acting at a point in one plane, and if everything is known about the system except two details, namely, *either* the magnitude of one of the forces and the direction of one of the forces, *or* the magnitudes of two of the forces, *or* the directions of two of the forces, we can in general . determine the two unknowns by the graphical method.

24. Ex. 1. *Find the resultant of forces of* 4, 5, 6 *pounds' weight acting at a point in one plane, the angle between the first two forces being* 37°, *and between the first and the third a right angle measured in the same direction.*

Take any straight line OH, and make the angles HOK, HOL equal to 37° and 90° respectively.

With any suitable scale, draw AB of length 4 units in the direction of OH, BC of length 5 units in the direction of OK, CD of length 6 units in the direction of OL.

Then AD represents the resultant, and a straight line OR, drawn through O parallel to AD, is its line of action.

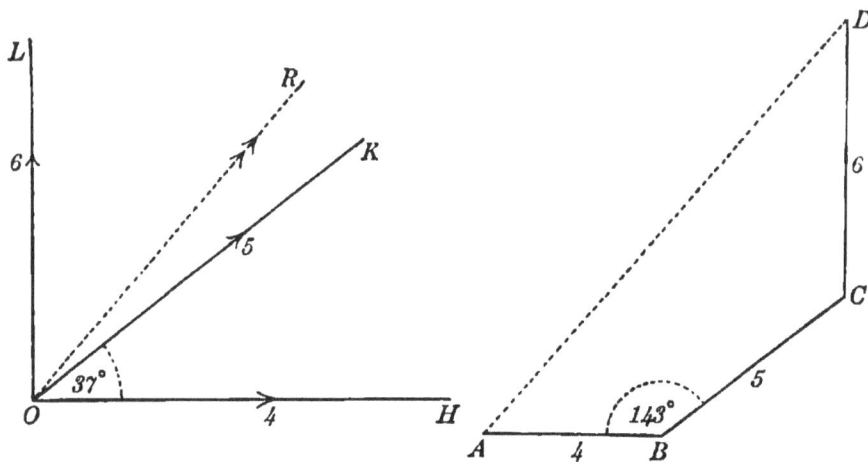

FIG. 24.   FIG. 24 a.

On measurement, we find that AD is of length 12 units, and that the angle BAD is $48\frac{1}{2}°$.

Hence, the resultant is 12 pounds' weight in a direction making $48\frac{1}{2}°$ with the first force.

25. Ex. 2. *Find the magnitudes of the forces P, Q, in order that the system of forces, represented in figure 25, may be in equilibrium.*

Draw AB of length 4 units in the direction of the force marked 4, BC of length 2 units in the direction of the force marked 2, CD of length 3 units in the direction of the force marked 3.

Through D and A draw straight lines parallel to the lines of action of the forces P and Q respectively, to meet in E.   Then DE and EA represent P and Q respectively.

On measuring DE, EA we find that
$$\begin{cases} P = 5\cdot8, \\ Q = 5. \end{cases}$$

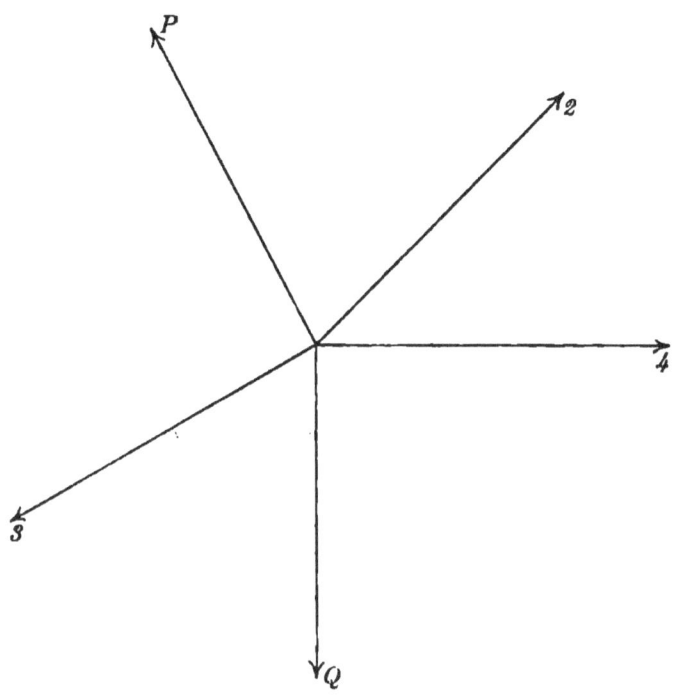

Fig. 25.

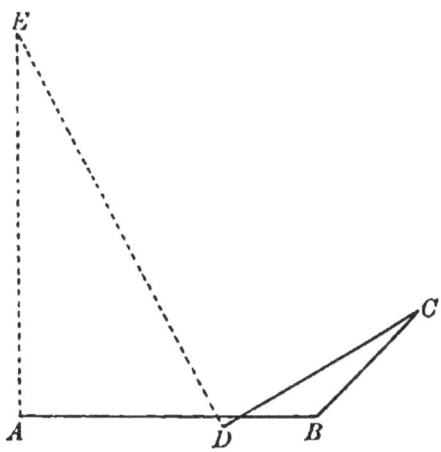

Fig. 25 a.

26. Ex. 3. *Two equal forces P are in equilibrium with two equal forces Q, all four being in one plane and acting at the same point. Prove that* either (i.) *the two forces P are in opposite directions and the two forces Q in opposite directions,* or (ii.) *the bisector of the angle between the two forces P is in the same straight line as the bisector of the angle between the two forces Q.*

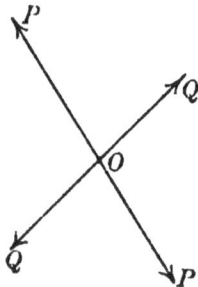

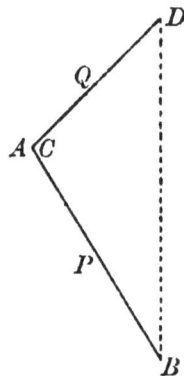

FIG. 26.   FIG. 26 a.

Let a force polygon be constructed taking the forces in the order *P, P, Q, Q*. This will be a quadrilateral *ABCD* in which *AB, BC* are each of length *P* units, and *CD, DA* each of length *Q* units. Hence the triangles *DAB, DCB* are equal in all respects, so that *DA* and *DC* are equally inclined to *DB*, and *AB, CB* are equally inclined to the same line.

(i.) Let *A* and *C* be on the same side of *DB*.

In this case *A* and *C* must coincide. Therefore the forces represented by *AB, BC* are in opposite directions, and the forces represented by *CD, DA* are in opposite directions.

(ii.) Let A and C be on opposite sides of DB.

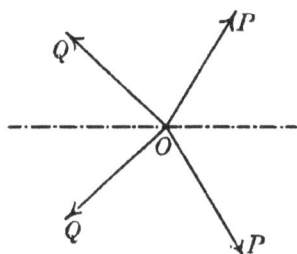

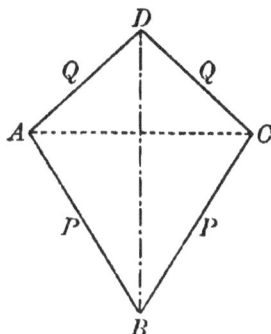

FIG. 27.   FIG. 27 a.

In this case a straight line parallel to AC is the bisector of the angle between the forces P, and also of the angle between the forces Q.

27. Ex. 4. *Find a point P, within a quadrilateral ABCD, such that the forces represented by PA, PB, PC, PD may be in equilibrium.*

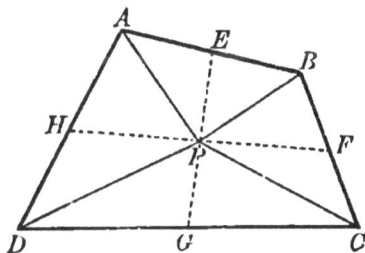

FIG. 28.

Find E and G the middle points of AB and CD respectively. Then the forces represented by PA, PB have for their resultant a force represented by twice PE; and the forces represented by PC, PD have for their resultant a force represented by twice PG.

Hence, for equilibrium, it is necessary and sufficient that P should be the middle point of EG.

D.S.   C

Hence, take P at the middle point of EG. Then we know that the forces represented by PA, PB, PC, PD are in equilibrium, and therefore, in the same way as above, P must be the middle point of the straight line joining F and H, the middle points of BC and DA respectively.

Thus, we have here an independent proof of the geometrical property of the quadrilateral, that the straight lines joining the middle points of opposite sides bisect one another.

EXAMPLES II.

1. If the side of a regular hexagon $ABCDEF$ represents a force of 100 pounds' weight, find the magnitudes of the forces represented by the straight lines AE, AD, FB; and, supposing them to act at a point, determine the magnitude and direction of the resultant of the three forces.

2. If a straight line AB represents a force equal to the weight of 1 pound, construct a line which shall represent a force equal to the weight of $3\sqrt{2}$ pounds.

A, B, C, D are the angular points of a square taken one way round, and forces represented in direction by the lines AB, BD, DA, and AC, and in magnitude by the numbers 1, $2\sqrt{2}$, 3 and $\sqrt{2}$, act at a point; find their resultant.

3. Forces 1, 2, 3 and $2\sqrt{2}$ act at a point in the directions of the sides AB, BC, CD and the diagonal DB of a square $ABCD$ respectively; determine their resultant.

4. $ABCD$ is a square; find the resultant of the forces represented by the straight lines AB, AC, and AD.

5. OA, OB, OC are three straight lines inclined at angles of $120°$ to one another; a force $3P$ acts from A towards O, a force $4P$ from O towards B, and a force $5P$ from O towards C. Determine the magnitude and direction of the resultant of the three forces.

6. Forces of 2, 3, x pounds' weight act at a point in one plane, the middle one being inclined to each of the others at an angle of 60°. If the force of 2 pounds' weight is removed, the resultant is of the same magnitude as before. Find x.

7. The triangle ABC has its sides BC, CA, AB of lengths 8, 12, 15 inches respectively. A particle is acted upon by forces of 2, 3, 1 pounds' weight parallel to and in the direction of BC, CA, AB respectively. Find the resultant.

8. Find the resultant of forces 4, 2, 5, 3 acting at a point in one plane, the angles between 4 and 2, 2 and 5, 5 and 3 being 90°, 30°, 120° respectively, and all angles being taken the same way round.

9. Forces 3, P, 5, 2, Q act at a point in one plane, the angles between 3 and P, P and 5, 5 and 2, 2 and Q being 90°, 60°, 60°, 90° respectively, all taken the same way round. Find P and Q in order that the system may be in equilibrium.

10. Find the resultant of forces 3, 5, 2, 4 acting at a point in one plane, the angles between 3 and 5, 5 and 2, 2 and 4 being 90°, 60°, 90° respectively, all taken the same way round.

11. Forces 2, P, 1, Q, 3 act at a point in one plane, the angles between 2 and P, P and 1, 1 and Q, Q and 3 being 30°, 90°, 30°, 90° respectively, all taken the same way round. Find P and Q in order that the system may be in equilibrium.

12. Find the resultant of forces 5, 3, 2 pounds' weight acting in one plane at a point, the middle one being inclined to each of the others at an angle of 60°.

13. Find the resultant of forces 11, 8, 3 pounds' weight acting in one plane at a point, the angle between each pair being 120°.

14. $ABCDEF$ is a regular hexagon. Upon a particle at A forces of 6, 8, 9, 8, 6 pounds' weight act in the directions AB, AC, AD, EA, AF respectively. Find their resultant in magnitude and direction.

15. $ABCDEF$ is a regular hexagon. Upon a particle at A forces of 12, 17, 6, 2, x pounds' weight act in directions AB, AC, AD, AE, AF respectively. Find x in order that the resultant may be in the direction AC.

16. ABC is an equilateral triangle, and D is the middle point of BC. Find the magnitude and direction of the resultant of the following three forces acting at A : 3 pounds' weight in direction AB, 2 pounds' weight in direction DA, 4 pounds' weight in direction AC.

17. $OABC$ is a square, and D is a point in AB such that AD is $\frac{3}{4}$ of AB. Find the magnitude and direction of the resultant of forces of 4, 2, 3 pounds' weight acting at O in directions OA, OD, OC respectively.

18. $OABC$ is a square, each side of which is 1 foot in length. D is a point in AB 5 inches from A, and E is a point in BC 3 inches from B. Find the magnitude and direction of the resultant of the following system of forces acting at O : 45 pounds' weight along OA, 65 pounds' weight along OD, 35 pounds' weight along EO, 66 pounds' weight along OC.

19. Let O be the position of a particle, and OA a straight line drawn through O. Find the magnitude and direction of the resultant of forces of 10, 18, 20, 16 pounds' weight acting on the particle, when their directions make with OA angles of 0°, 30°, 90°, 135° respectively, all measured in the same sense.

20. Forces of magnitudes 3, 4, and 5 act at a point O in directions lying in one plane, and making angles of 15°, 60°, and 135° respectively, with a line OA in the same plane. Find the magnitude of the resultant.

21. Forces of 3, 4, and 6 pounds' weight make angles of 90°, 60°, and 30° respectively with a force of 2 pounds' weight (the angles being measured in the same direction). Find the magnitude of the resultant, and the angle its direction makes with the force of 2 pounds' weight.

22. $ABCDEF$ is a regular hexagon ; forces of 1, P, 2, Q, 6 pounds' weight respectively act along the lines AB, AC, AD, AE, AF. Find the value of P in order that the resultant of the system may be along AE.   ·

23. A particle is acted upon by three forces of given magnitudes ; show how these forces must be arranged so as, if possible,

to produce equilibrium, and determine the angle between the last two forces, when the measures of the forces are

(i.)　6, 11, 18 ;
(ii.)　1, 1, $\sqrt{2}$;
(iii.)　17, 15, 8 ;
(iv.)　13, 15, 7 ;
(v.)　13, 15, 8 ;
(vi.)　13, 8, 7.

24. Two equal forces are in equilibrium with a third force which is fully known. If the direction of one of the equal forces be known, show how to determine the direction of the other and the magnitude of each.

25. Find a point P within a triangle ABC, so that the forces represented by PA, PB, PC may be in equilibrium. Make use of this to prove the geometrical theorem, that the three medians of a triangle are concurrent; and that the distance of their point of concurrence from a corner is two-thirds of the length of the median along which it is measured.

26. Extend Art. 10 to include the case of any number of forces acting at a point.

27. Four forces in equilibrium, acting at a point, are represented in magnitude and direction by AB, CD, AD, CB. Show that A, B, C, D must be the angular points of a parallelogram.

28. A number of forces, acting at a point in one plane, are in equilibrium. If one of them be turned about its point of application through a given angle, show how to find the resultant of the system, and, if the inclination of the force continue to alter, show that the inclination of the resultant alters by half the amount.

29. Three forces, whose measures are P, Q, X, are in equilibrium when acting at a point; the first force is given in magnitude and position, the second in magnitude only, the third in direction only, making an angle θ with the direction of the first. Show how to determine the direction of the second and the measure of the third. Show that there are generally two solutions, and that, if $P > Q$, there are limits to the angle θ, beyond which the question is impossible of solution.

As an example, take the case in which $P = 15$, $Q = 13$, $\theta = 120°$.

30. In the preceding question, show that the product of the two values of X is equal to the difference between the squares of P and Q.

31. Three forces P, Q, R act at a point O, and are in equilibrium. A circle through O cuts their lines of action in p, q, r respectively. Prove that $P : Q : R = qr : rp : pq$.

32. Three forces, acting at a point, are in equilibrium. Show that if a triangle be formed by drawing straight lines perpendicular to the directions of the forces, its sides will be proportional to the forces to which they are respectively perpendicular.

33. Three forces P, Q, R, in equilibrium, act along the lines OA, OB, OC, where O is the orthocentre of the triangle ABC. Prove that $P : Q : R = BC : CA : AB$.

34. I is the centre of the circle inscribed in the triangle ABC. Three forces P, Q, R, in equilibrium, act along the lines IA, IB, IC respectively. Prove that $P : Q : R = BC : CE : EB$, where E is the centre of the circle which touches BC, AB produced and AC produced.

35. E is the centre of the circle which touches BC, AB produced and AC produced. Three forces P, Q, R, in equilibrium, act along the lines AE, EB, EC respectively. Prove that $P : Q : R = BC : CI : IB$, where I is the centre of the circle inscribed in the triangle ABC.

36. A, B, C are three points on the lines of action of three forces P, Q, R respectively, which act at O and are in equilibrium. Prove that if $P : Q : R = BC : CA : AB$, then O is either the orthocentre of the triangle ABC, or it is some point on the circle which passes through A, B, C.

37. In the preceding example, prove that if $P : Q : R = BC : AB : CA$, then either BC is a common tangent to the circles BOA, COA, or O coincides with A; and that, in each case, the line of action of P passes through the middle point of BC.

CHAPTER III.

EQUILIBRIUM OF FINE LIGHT STRINGS IN A STATE OF TENSION.

28. By a *light* string we mean one whose weight is inappreciable. By a *fine* string we mean one whose thickness is inappreciable.

29. When a string in a state of tension has taken up a position of equilibrium, we may treat any portion of it as a rigid body at rest under the influences of the forces which act externally upon that portion.

This is a particular case of the following important general principle: The conditions of equilibrium of a body not rigid are the same as those of an ideally rigid body with these additions:—(i.) Every portion into which the body can be conceived to be divided must be in equilibrium under the external forces which act upon that portion considered as a rigid body. (ii.) The external forces acting upon the body must not be such as to induce internal actions within the body sufficient to break or fracture it.

The first of these additional conditions enables us to find, when necessary, the internal actions at any point within a body; the second we generally ignore in the elementary statics, as we generally assume that

the material system under consideration is strong
enough to bear any strain to which it may be subjected.
In practical applications this, of course, has to be taken
into account.

30. · Now let AB be a portion of a fine light string
in equilibrium in a state of tension, and suppose that
between the points A and B the string is *quite free*.

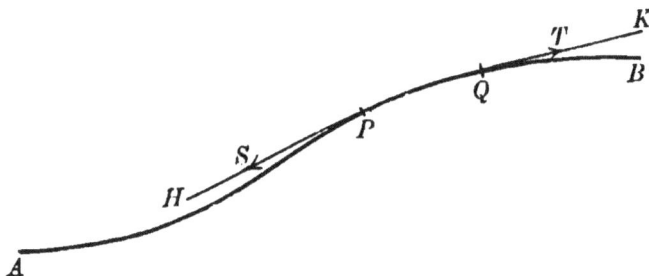

FIG. 29.

Take *any* two points P and Q of the string between
A and B, and consider the equilibrium of the portion
PQ of the string.

The fibres at P are in a state of tension, so that
the adjoining piece of string AP is pulling upon · the
piece PQ at P with a force S in the direction of
the tangent PH.

Similarly, the fibres at Q are in a state of tension,
so that the adjoining piece of string QB is pulling
upon the piece PQ at Q with a force T in the direction
of the tangent QK.

Now we *assume* that between P and Q the string
is strong enough to bear all strain to which it is
subjected, and that if a *rigid body* of the same size
and shape as PQ were substituted for the string PQ,
it would be in equilibrium under the same external
forces.

But, if a rigid body is in equilibrium under the influence of two forces, those forces must be equal and opposite and act in the same straight line.

$\therefore$ $S = T$, and HP and QK are parts of the same straight line.

Also, as P and Q were taken to be *any* two points between A and B, we see that the magnitude of the tension is the same at every point between A and B, and that the tangents at all points of AB are in one and the same straight line. Hence the portion AB must be straight.

Thus, if a fine light string is in equilibrium in a state of tension, every *free* portion of it is straight, and the tension is the same at every point of such a portion.

31. In particular, let a fine light string AB rest in equilibrium with one extremity A attached at a fixed point, and with a force whose measure is F applied at B in a fixed direction.

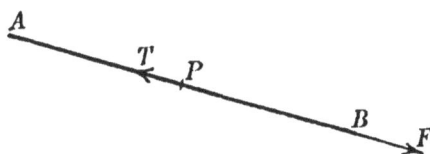

Fig. 30.

By Art. 30, we see that AB must be straight, and the tension at any point is the same throughout the string. Let T be the measure of this tension.

Take any point P of the string, and consider the equilibrium of the portion PB as a rigid body.

The external forces acting upon it are F in the

fixed direction and T in the direction PA. These must be equal and opposite.

$$\therefore \quad T = F,$$

and AB is in the direction of the force F.

Thus the string takes up a position of equilibrium such that AB is in the direction of the force applied, and the fibres of the string pull both ways at every point with a force equal in magnitude to the force applied.

If a heavy mass is attached at B, then AB takes up the vertical position with B below A, and the tension of the string at every point is equal to the weight of the mass supported.

32. Suppose now that a number of fine light strings OA, OB, OC, OD, in a state of tension, are knotted together at the point O, and that they have taken up some position of equilibrium.

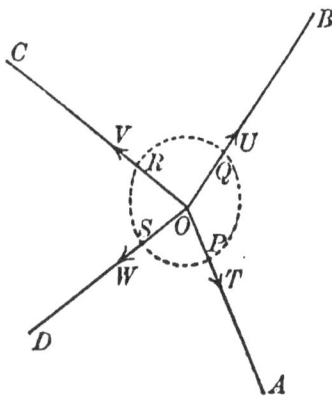

FIG. 31.

The tension of the string OA is the same at every point. Let its measure be T.

Similarly let U, V, W be the measures of the tensions of OB, OC, OD respectively.

Consider the equilibrium of the portion of string in the immediate neighbourhood of O. We have drawn a closed curve round O cutting off the portions OP, OQ, OR, OS from the strings. The portion of string within this closed curve we treat as a rigid body in the manner explained above.

The forces acting externally upon this portion are as follows:

At P it is being pulled in direction PA with a force T.

At Q „   „ QB „ U.

At R „   „ RC „ V.

At S „   „ SD „ W.

The lines of action of these forces meet at the point O, and we may treat them as though they all acted at O. Hence the consideration of the equilibrium of this portion of matter round O comes under the case of the consideration of the equilibrium of a number of forces acting at a point. The force polygon will have its sides parallel to the lines OA, OB, OC, OD, and the arrows must go one way round.

The consideration of the equilibrium of the portion of the system in the neighbourhood of O is briefly described as considering the equilibrium of O, and the force polygon is called the force polygon for the point O.

33. *An endless fine light string LMN is in a given position, in the form of a triangle, with the point N fixed. To the point L is applied a known force, whose measure is P, in a given direction LH outwards from the triangle. It is required to find the measure of a force which must be applied at M in a given direction MK, in order that the system may be in equilibrium,*

*and also to determine the tensions in the different parts
of the string and the force of constraint at N.*

Let X be the unknown measure of the force applied
at M, and let NV be the unknown direction, Y the
unknown measure, of the constraint at N.

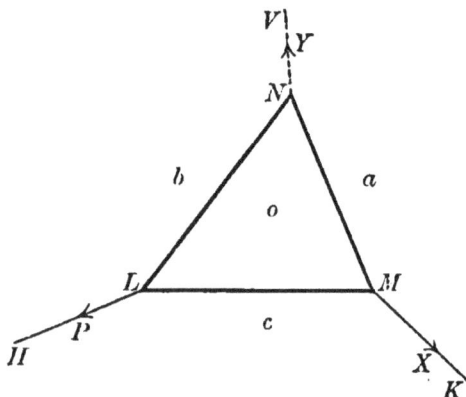
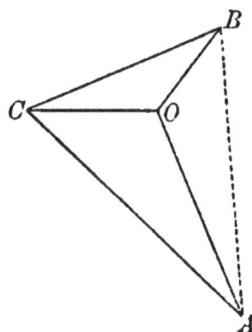

FIG. 32.   FIG. 32 a.

We will make use of the notation suggested in
Art. 4, and the student will at once see the advantage
of doing so. The lines of the space diagram divide it
into four parts which we letter o, a, b, c as in the
figure. We have now no further use for the letters
L, M, N, H, K, V in the diagram. The straight lines
LH, LM, etc., we now call bc, oc, etc. The point L,
where the spaces b, c, o meet, we now call bco, etc.

Take BC of length P units in the direction of the
given force P, *i.e.* parallel to bc, and draw BO, CO
parallel to bo, co respectively, meeting in O. Then
$BCOB$ (this way round) is the triangle of forces for
the point bco, so that CO and OB represent the pulls
of the strings co and ob respectively upon the point bco.

Now consider the equilibrium of the point oca. The pull of the string co at this point is represented by OC. Draw CA, OA parallel to ca, oa respectively, meeting in A. Then $OCAO$ (this way round) is the triangle of forces for the point oca, so that AO represents the pull of the string ao at the point oca, and CA represents the force applied at M.

Now consider the equilibrium of the point oab. The pulls of the strings bo, ao at this point are represented by BO, OA respectively. Therefore the remaining force of constraint at N, which balances these two, must be represented by AB, the triangle of forces for the point oab being $OABO$ (this way round).

Measuring the lines of the force diagram, we have X, Y and the tensions of the three parts of the string, and AB gives us the direction also of the force of constraint.

34. Ex. 1. *The fine string $ABCDE$, of length 3 feet, has its extremities attached to the two points A and E, situated 18 inches apart in a horizontal line. Another fine string, of length 10 inches, connects the points B and D, situated 5 inches each from A and E respectively, and to the middle point C of the first string is attached a mass of 24 pounds. The whole is allowed to take up a symmetrical position of equilibrium. Find the tension of each portion of the string.*

There is no difficulty in constructing the space diagram to scale; this the student should do for himself. We then mark the portions of the space diagram, as in the figure, with the letters o, h, k, l.

With any suitable scale, draw HK 24 units of length vertically downwards to represent the tension of the string hk, which is equal to the weight of 24 pounds. Draw HO, KO parallel to the strings ho, ko respectively, to meet in O.

Then $HKOH$ (this way round) is the force triangle for the point hko. Thus KO represents the pull of

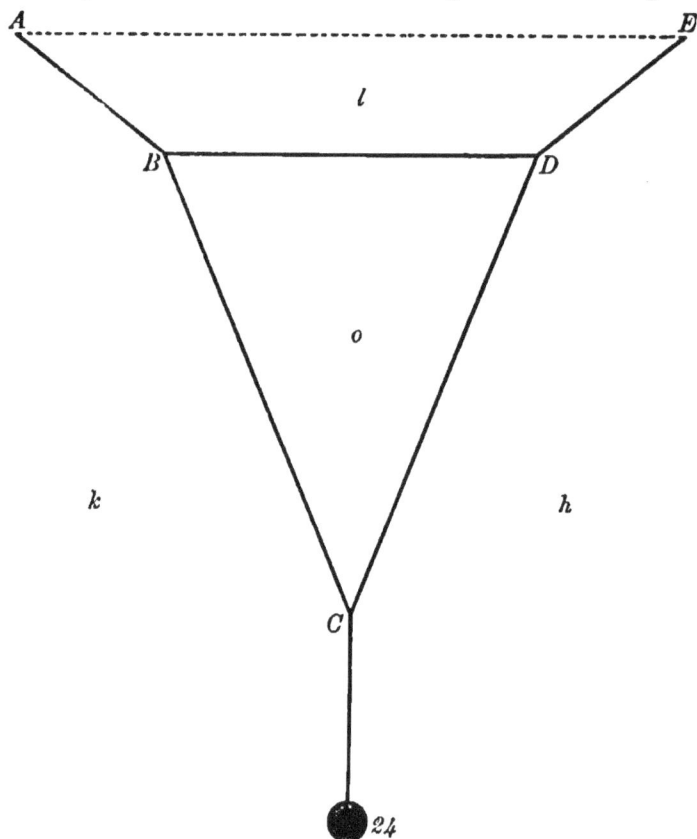

FIG. 33.

the string ko at C; $\therefore$ OK represents the pull of the same string at B.

Draw OL, KL parallel to the strings ol, kl respectively, meeting in L. Then $OKLO$ (this way round) is the force triangle for the point okl. Thus LO represents the pull of the string lo at B; $\therefore$ OL represents the pull of the same string at D. Again, since OH represents the pull of the string oh at C, HO represents the pull of the same string at D. Hence we see that HO, OL represent the pulls of the two strings ho, ol at D; therefore, by joining H, L we complete the force

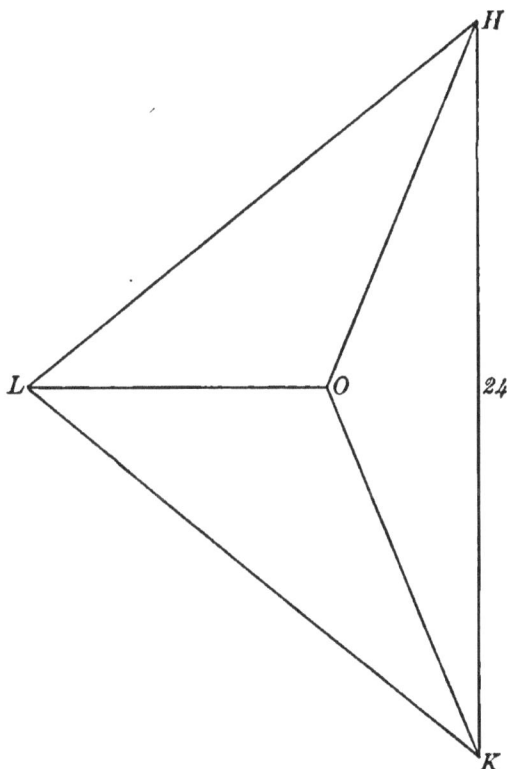

FIG. 33 a.

diagram, and the line LH must be parallel to the string lh, and must represent its tension.

If we consider the equilibrium of the strings which form the triangle BDC as one rigid body, the force diagram for the system is the triangle HKL, the lines HK,

KL, LH representing the external forces which act upon the triangle CBD at the points C, B, D respectively.

On measuring the lines of the force diagram, we have $LK = 20 = LH$, $OK = 13 = OH$, $LO = 11$.

∴ The tensions of the strings BC, CD are each 13 pounds' weight, of AB, DE each 20 pounds' weight, and of BD 11 pounds' weight.

35. Ex. 2. *A fine light string ACB is placed on a smooth horizontal table, and has its extremities fastened to two given fixed points A, B. A force is applied in the plane of the table to the string at the point C, which is at given distances measured along the string from A and B. It is required to find the conditions under which the string will rest in equilibrium with both portions in a state of tension, and, when the applied force is given satisfying these conditions, to determine the tensions of each portion of the string.*

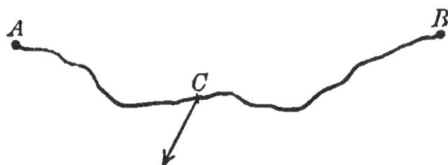

FIG. 34.

At the outset we do not know the position of equilibrium, but, in any position of equilibrium in which both portions of the string are in a state of tension, each of those portions will be straight. Now we are given the lengths of the portions AC, BC, and also the positions of the points A, B. Hence the point C will take up one or other of two positions C_1, C_2, which we can find, situated symmetrically on opposite sides of AB.

Let HK represent the applied force. Having found the positions of the points C_1, C_2, draw Ha_1, Hb_1, Ha_2, Hb_2 in the directions of the lines AC_1, BC_1, AC_2, BC_2 respectively.

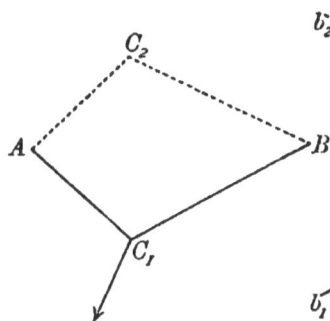 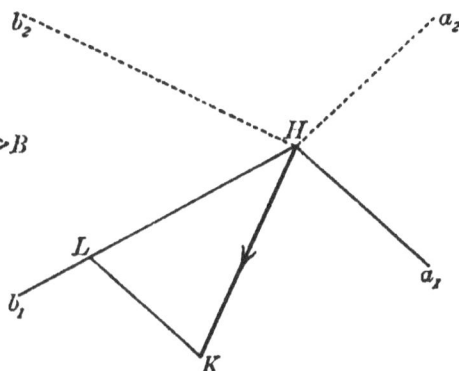

FIG. 35.   FIG. 35 a.

If AC_1B is the position of equilibrium, considering the equilibrium of the portion of string in the neighbourhood of the point C_1, we see that the applied force will have to balance two forces in the directions C_1A, C_1B, and therefore HK must lie within the angle a_1Hb_1. Similarly, for the position AC_2B, HK must lie within the angle a_2Hb_2. Thus, for the string to rest with both portions in a state of tension, the applied force must be between the directions Ha_1 and Hb_1, or between Ha_2 and Hb_2. If there is no limit to the possible tension of the string, there is no further limitation upon the magnitude or direction of the applied force.

If the line HK be given, within, say, the angle a_1Hb_1, then, drawing KL parallel to a_1H to meet Hb_1 in L, $HKLH$ (this way round) is the triangle of

D.S.   D

forces for the point C, and the position of equilibrium is AC_1B.

. If the string can stand at every point tension up to, but not beyond, a certain given value, describe with centre H, and radius representing this maximum tension, a circle intersecting Ha_1, Hb_1, Ha_2, Hb_2 in a_1, b_1, a_2, b_2 respectively, and complete the parallelograms $b_1Ha_1K_1$, $b_2Ha_2K_2$.

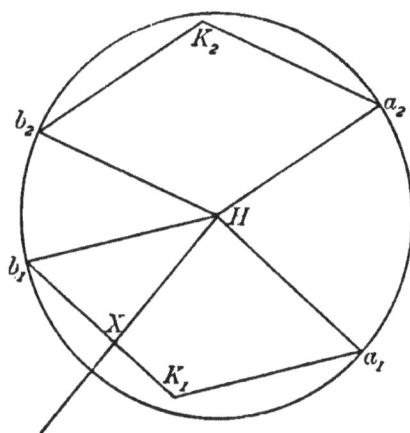

FIG. 36.

Then the point K must lie within one or other of these two parallelograms.

If it be required to find the greatest force which can be applied in any given direction without breaking the string, we have merely to find the point X where the straight line drawn from H in the given direction meets $a_1K_1b_1$ or $a_2K_2b_2$. Then HX represents the force required.

36. Ex. 3. *If, in the preceding example, the line of action of the applied force passes through D, the middle point of AB, then the tensions of the two portions of the string are proportional to their lengths.*

Draw through A a straight line parallel to CB to meet CD produced in E. Then the triangles ADE, BDC are similar, and, as $AD=DB$, it follows that $AE=CB$.

Now the triangle CAE has its sides parallel to the three forces which keep the portion of string at C in

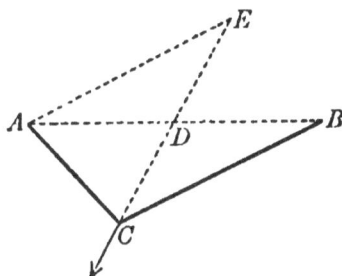

FIG. 37.

equilibrium; therefore its sides are proportional to the forces to which they are respectively parallel.

$\therefore$ the tensions of the strings CA, CB are proportional to CA, AE, that is to CA, CB.

EXAMPLES III.

1. A mass of 24 pounds is supported by two fine light strings inclined at angles of 15° and 60° with the vertical. Find the tensions of the strings.

2. A and B are two fixed points distant 8 feet apart in a horizontal line. Two fine light strings AC, BC, of lengths 5 and 7 feet respectively, support a mass at C. Compare their tensions.

3. BAC is a fine light string, of length 14 inches, attached at its extremities to two points B and C, situated 10 inches apart. At the point A, 6 inches from B, is knotted another string AD, which is pulled with a force equal to the weight of 20 pounds. If DA produced passes through E, the middle point of BC, find the tensions in the strings BA and AC.

4. Two fine light strings AB, BC, each of length 5 feet, are knotted together at B and attached at their other extremities to fixed points A and C, situated 8 feet apart in a horizontal line. A mass of 180 pounds is supported by another fine light string attached at B. Find the tension in each of the strings AB, BC.

5. Two fine light strings AB, BC, of lengths 13 and 15 inches respectively, are knotted together at B, their other extremities being attached to two points A and C, situated 14 inches apart in a horizontal line. A third string, attached at B, supports a mass weighing one cwt., and the whole is allowed to hang freely. Find, in pounds' weight, the tensions of the strings.

6. A mass of 300 pounds is supported by two fine light strings, of lengths 17 and 26 inches respectively, attached to the same point of the mass, the other extremities of the strings being respectively attached to two points, situated 25 inches apart in a horizontal line. Find the tensions of the strings.

7. A mass of 42 pounds is supported by two strings AC, BC, of lengths 17 and 25 inches respectively, attached to two points A, B situated 28 inches apart in a horizontal line. Find the tensions of the strings.

8. Four fine light strings, each of length 5 inches, are knotted together to form a rhombus $ABCD$, which is suspended from A. A mass of 80 pounds is attached at C, and B, D are kept 6 inches apart in a horizontal line by two equal and opposite forces P acting at B and D. Represent the forces acting upon each of the knots B, C by the sides of a triangle, and find the magnitude of P and the tension of the string.

9. Three fine light strings are knotted together to form a triangle ABC, the strings AB, BC, CA being of lengths 8, 5, 5 inches respectively. If a mass of 30 pounds is suspended from C, and the whole is supported, with AB horizontal, by two forces applied at A and B in directions making $22\frac{1}{2}°$ with the horizontal, find the tension of each portion of the string and the magnitude of each of the applied forces.

10. Three fine light strings are knotted together to form a triangle ABC, the strings AB, BC, CA being of lengths 28, 25,

17 inches respectively. The point A is fixed, and a mass of 84 pounds is suspended from C. If the triangle and mass are supported, with AB horizontal, by a force applied at B in direction perpendicular to AC, find the magnitude of this force, the tension of each portion of the string, and the magnitude and direction of the action at A.

11. A fine light string $ABCDE$, of length $5\frac{1}{2}$ feet, has its extremities attached to two points A and E, situated 4 feet apart in a horizontal line. Another piece of string, of length 2 feet, connects the points B and D, situated 13 inches each from A and E respectively ; and to the middle point C of the first string is attached a mass weighing 40 pounds. The whole is allowed to take up a symmetrical position of equilibrium. Find the tension of BD.

12. A fine light string ABC supports a mass, of given weight, at C, and is attached to a fixed point at A. To the point B of the string, situated at a given distance from A measured along the string, a given force is applied in a given direction. Show how to find the position of equilibrium, and the tension of AB.

For example, if the mass supported is 40 pounds, and the given force is equal to the weight of 15 pounds, and acts in a direction of 60° with the upward vertical direction, find the tension of AB and the vertical distance of B below A, the distance AB being 14 inches.

13. In the preceding example, show how to find the magnitude and direction of the smallest force which will cause the string to rest in a given position.

14. A fine light string of given length has its extremities attached to two given fixed points. Show how to find the greatest load that can be applied to a given point of the string without breaking it, supposing that string can bear any tension up to a certain given value.

CHAPTER IV.

EQUILIBRIUM OF FINE LIGHT RODS, FREE EXCEPT AT THEIR EXTREMITIES.

37. By a *light* rod we mean one whose weight is inappreciable. By a *fine* rod we mean one whose thickness is inappreciable.

38. We shall in this chapter confine our attention to *straight* rods of no appreciable weight or thickness, which are in equilibrium under forces applied only at their extremities, so that each rod is quite free throughout its length.

If a framework of rods is in equilibrium, each individual rod must be in equilibrium, and each part of the structure must be in equilibrium considered as a rigid body, whether such part consists of a certain number of the rods which make up the structure, or of parts of the rods themselves.

In considering the equilibrium of a system of such rods jointed together at their extremities to form a framework, as we neglect the thicknesses of the rods, so also we shall neglect the sizes of the hinges. We shall suppose that all hinges are smooth, and that the effect of a hinge upon a rod is to compel that extremity of the rod to remain in a definite position by means of a direct push or pull applied at that point. The constraint is a self-adjusting force, and accommodates itself to prevent, if possible, the extremity of the rod from getting away from the part of the structure to which it is attached. It is of any magnitude, and acts in any direction necessary to preserve equilibrium, but must act through the extremity of the rod, which extremity is here treated as a mere point.

39. .Let *AB* represent a rod in equilibrium, being jointed freely at *A* and *B* to the adjoining parts of the structure. Suppose also that its weight is inappreciable, and that it is quite free between *A* and *B*, so that the only external forces that act upon it are applied at *A* and *B*.

FIG. 38.

FIG. 39.

The forces that act upon it at *A* are equivalent to a single force acting {upon it at *A*. So also the forces at *B* are equivalent to a single force at *B*. Thus the forces acting externally upon the rod are equivalent to two forces, one acting at *A* and the other at *B*. These two must be equal and opposite and act along the same straight line. Therefore AB must be the line of action of both forces. Also the resultant actions at A and B must be *either* (i.) both inwards, in which case the rod is in a state of compression, and its effect is to keep the two parts of the structure at A and B apart from one another; such a rod is called a *strut*; or (ii.) both outwards, in which case the rod is in a state of tension, and its effect is to bind together the two parts of the structure at A and B; such a rod is called a *tie*.

In the case of a *strut*, the rod *pushes* the structure at each end in its own direction. In the case of a *tie*, the rod *pulls* the structure at each end in its own direction.

40. Let us consider the equilibrium of the portion AP of the rod AB, which is acted upon by forces at A and B only. Since the portion AP is in equilibrium,

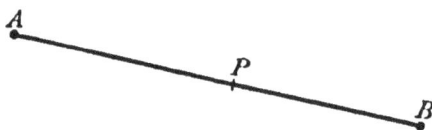

FIG. 40.

the action of the adjoining piece PB at P upon the portion AP, must be equal and opposite to and in the same straight line as the action at A. Therefore, in the case of a *strut*, the portion PB *presses* against the portion AP with a force equal and opposite to the resultant of the forces which act upon the rod at A. In the case of a *tie*, the portion PB *pulls* at the portion AP with a force equal and opposite to the resultant of the forces which act upon the rod at A.

In the one case the fibres at P are in a state of compression, in the other case in a state of tension. In *both* cases the action at P of one part upon the other is in the direction of the rod, and is of the same intensity at every point of the rod.

In the case of a *tie*, we might replace the rod by a string which would answer the purpose· theoretically just as well.

If a rod is *not* free between its extremities, or if it is of appreciable weight, then the action at each end is not necessarily in its own direction, and the internal strains may be different at every point of the rod.

41. The equilibrium of a *tie* is *stable*; for, if it be twisted a little out of its position, the external forces acting upon it at its extremities tend to restore it to

Fig. 41.

Fig. 42.

its original position. The equilibrium of a *strut* is *unstable*; for, if it be displaced, the external forces acting upon it at its extremities tend to twist it still further from its original position.

42. Now suppose that several rods, all coming under the case above described, are jointed together at a common extremity.

Let the rods OA, OB, OC, OD be freely jointed at O. Draw a closed curve round O, cutting off portions OP, OQ, OR, OS from the rods, and consider the equilibrium of the matter contained within this curve

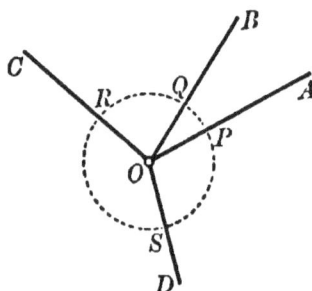

FIG. 43.

as a rigid body. Its weight is inappreciable, and the only forces acting externally upon it are the actions at P, Q, R, S which are along the lines (inwards or outwards) OA, OB, OC, OD respectively. Hence the consideration of the equilibrium of this portion of matter round O comes under the case of the consideration of the equilibrium of a number of forces acting at a point. The force polygon for the system under consideration will consist of a polygon having its sides parallel to the lines OA, OB, OC, 'OD, and the arrows must go one way round. The direction of the arrow decides in each case, not already known, whether the rod is a *strut* or a *tie*.

The consideration of the equilibrium of the portion of the system in the neighbourhood of O is briefly described as considering the equilibrium of O, and the force polygon is called the force polygon for the point O.

43. *Four fine light rods are freely jointed at their extremities to form a quadrilateral, which is stiffened by another fine light rod connecting two opposite joints. It is required to consider the equilibrium of the framework under the influence of two forces applied at the remaining two joints.*

Let the spaces inside the quadrilateral be denoted by o_1, o_2, and the spaces outside by a and b, as indicated in the diagram.

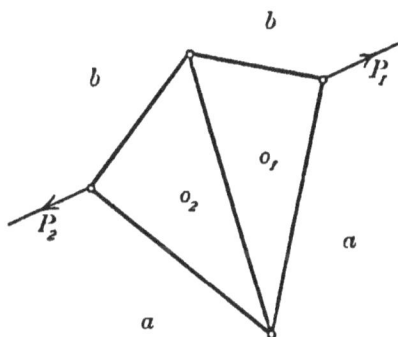

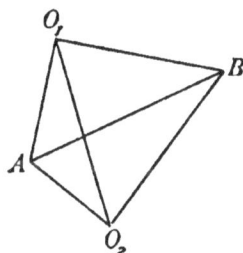

FIG. 44.   FIG. 44 a.

Let P_1, P_2 be the measures of the forces applied at the joints abo_1, bao_2 respectively.

We will first consider the equilibrium of the whole framework as one rigid body. It will be seen that the framework is not deformable and behaves as a rigid body. As the stresses in the rods can be of any magnitude and in either direction, it is necessary and sufficient for equilibrium that the forces acting externally upon the framework should form a system in equilibrium. In order to ensure, therefore, the equilibrium of the framework, it is necessary and sufficient that P_1 and P_2 should form a system in equilibrium. Thus, P_1 and P_2 must be equal and in opposite directions along the same straight line, either both inwards or both outwards. If P_1 and P_2 satisfy these conditions, the framework is in equilibrium, and we can determine the stresses in the rods.

Taking P_1 and P_2 as both outwards, the force diagram for the point abo_1 is a triangle ABO_1A (this way round), in which AB represents P_1 and AO_1, BO_1 are parallel to the rods ao_1, bo_1 respectively. The triangle of forces for the point o_1bo_2 is $O_1BO_2O_1$ (this way round), in which BO_2, O_1O_2 are parallel to the rods bo_2, o_1o_2 respectively. Now considering the joint ao_1o_2, we see that two of the forces acting upon it are represented by AO_1, O_1O_2. Hence, joining AO_2, the straight line AO_2 must be parallel to the rod ao_2, and the triangle of forces for the point ao_1o_2 is AO_1O_2A (this way round). Also the triangle of forces for the point o_2ba is O_2BAO_2 (this way round).

We see that the outside rods are *ties* and the cross rod a *strut*. We might replace the outside rods by strings.

If P_1 and P_2 both act inwards, it will be found that the outside rods are *struts* and the cross rod a *tie*. The force diagram will be the same as before, but the directions will in each case be the opposite way round. We might now replace the cross rod by a string.

In the above, the two equal and opposite forces P_1 and P_2 may be applied by means of another fine light rod, in a state of stress, connecting the joints abo_1 and bao_2. For instance, in a quadrilateral framework, if there are two diagonal ties and no external forces, we can determine the stresses in all the members provided we know the stress in one of them.

44. Ex. 1. *A fine light rod HK, of length 9 inches, is capable of turning freely in a vertical plane about the point H, which is fixed. To the point K is attached a fine light string, supporting at its other extremity a mass of 40 pounds. Another fine light string, of length 7 inches, connects K with the fixed point L, situated 8 inches vertically above the point H. Find the tensions of the strings, the stress in the rod, and the action at H.*

In the position of equilibrium, the two strings are straight and the mass rests vertically below K. The data are sufficient to enable us to construct the space diagram to scale. This done, we mark the portions of the space diagram with the letters a, b, o, as indicated in the figure on the next page.

The rod is at rest under forces acting only at H and K. Therefore the stress in the rod is at every point in its own direction, and the action at H is in the line HK.

The tension of the vertical string we see at once is 40 pounds' weight. Hence, draw AB vertically downwards of length 40 units, and through A and B draw AO, BO parallel to ao, bo respectively, to meet in O. Then ABOA (this way round) is the triangle of forces for the equilibrium of the portion of matter in the neighbourhood of K. On measuring, we find that BO and OA are of lengths 45 and 35 units respectively, and the direction BO shows that the rod is a strut. Hence the thrust of the rod is 45 pounds' weight, the tension of KL is 35 pounds' weight, and the action at H is 45 pounds' weight in direction HK.

In this example we might have dispensed with the force diagram altogether. The triangle LHK has its sides parallel to the forces which act upon the portion of the system at K. Hence, on the scale in which

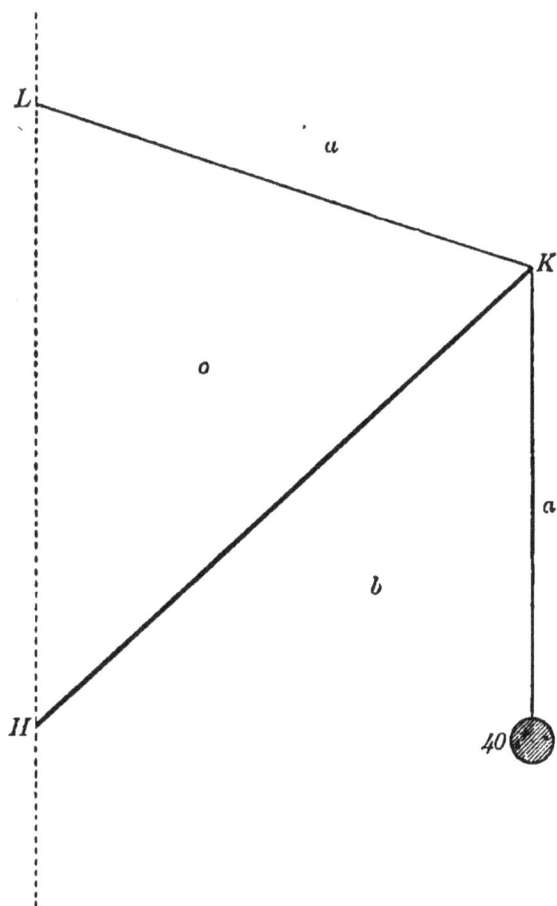

FIG. 45.

LH represents 40 pounds' weight, HK and KL represent the thrust of the rod and the tension of the string KL. This, of course, gives the same result as before.

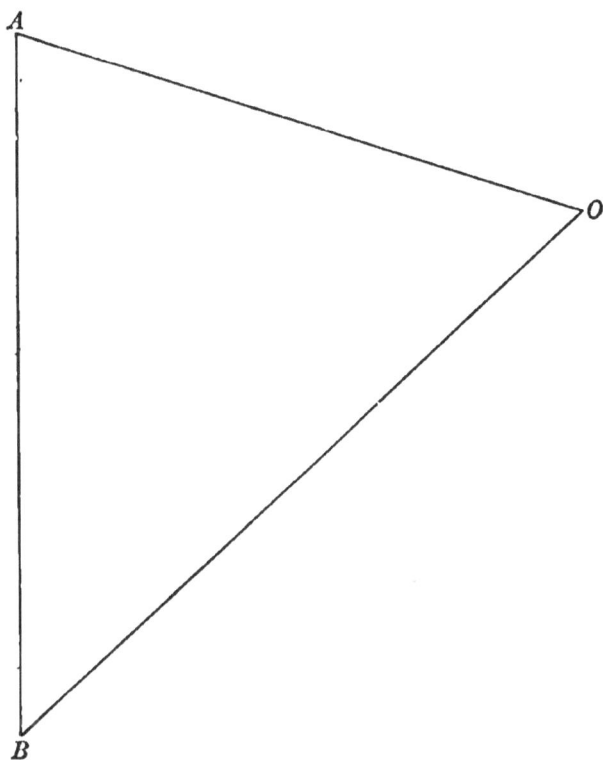

FIG. 45 a.

45. Ex.´2. *A fine light rod HK, of length* 56 *inches, is connected with a fixed point M by two fine light strings HM and KM of lengths* 39 *and* 25 *inches respectively. Another fine light string, of length* 112 *inches, has its extremities attached to the points H and K, and supports a mass of* 1 *cwt. at the point N, distant* 60 *inches along the string from H. The whole is allowed to rest in a vertical plane. Find the position of equilibrium, the tension of each portion of string, and the thrust in the rod.*

In the position of equilibrium the four strings are straight. The data are sufficient to enable us to con-

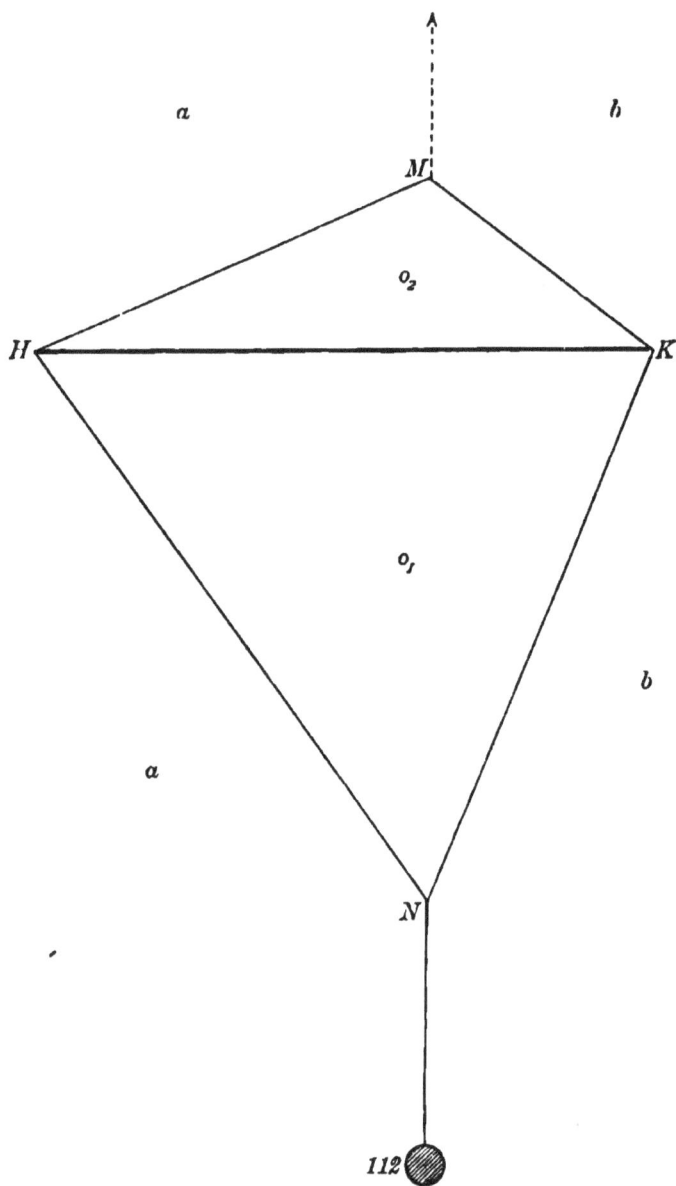

FIG. 46.

struct the *shape* of the quadrilateral $HMKN$, but not its position relatively to the vertical. We construct the space diagram to scale, and put in the vertical afterwards.

Considering the equilibrium of the four strings and the rod as one rigid body, we see that the force of constraint at M balances a force equal to the weight

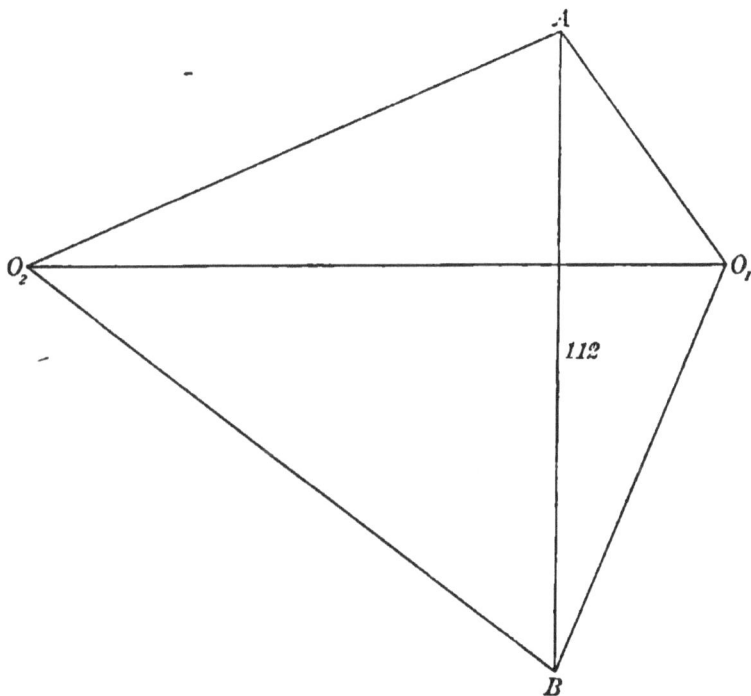

FIG. 46 *a*.

of 112 pounds acting vertically downwards through N. Hence the force of constraint at M must be equal to the weight of 112 pounds, acting vertically upwards, and the line MN must be vertical. Hence, joining MN, we have the vertical line through M, and thus determine the position of equilibrium. On measuring the angle

D.S.   E

between MN and HK we find that it is a right angle. Hence, in the position of equilibrium, HK is horizontal.

Having marked the parts of the space diagram with the letters a, b, o_1, o_2 as in the figure, we draw AB vertically downwards, and of length 112 units. The point O_1 is obtained by drawing through A and B straight lines parallel to the strings ao_1 and bo_1 respectively, the point. O_2, by drawing through O_1 and B straight lines parallel to o_1o_2 and bo_2 respectively; then, joining AO_2, we complete the force diagram.

On measuring the lines BO_1, O_1A, AO_2, O_2B, O_1O_2, we find that the tensions of the strings KN, NH, HM, MK, and the thrust of the rod are 78, 50, 104, 120, and 126 pounds' weight respectively.

46. Ex. 3. *Four fine light rods are smoothly jointed at their extremities to form a quadrilateral, which can be inscribed in a circle. The opposite joints are connected by two fine light strings in a state of tension. Prove that the thrusts in the rods and the tensions of the strings are proportional to the opposite sides and diagonals of the quadrilateral respectively.*

Let $ABCD$ be the framework, AC and BD being the diagonal ties.

Mark the line AB with the letters $c'd'$, placing one letter on each side of the line; AC with the letters $b'd'$, BC with the letters $a'd'$, etc.

Then the force diagram will be the quadrilateral $A'B'C'D'$, in which $A'B'$ is parallel to $a'b'$, $A'C'$ to $a'c$ $B'C'$ to $b'c'$, and so on.

We can now prove that the figure $A'B'C'D'$ is similar to the figure $ABCD$, the correspondence being shown

by using the same letters.   For,

angle $D'C'A'$ = angle DBA = angle DCA ;

also,   angle $D'A'C'$ = angle DBC = angle DAC ;

therefore the triangles $D'A'C'$ and DAC are similar. In the same way we can show that any other triangle of the figure $A'B'C'D'$ is similar to the corresponding triangle of the figure $ABCD$; therefore the two figures are similar.

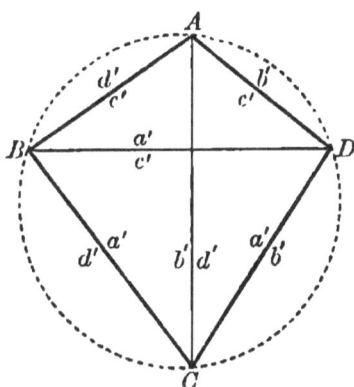
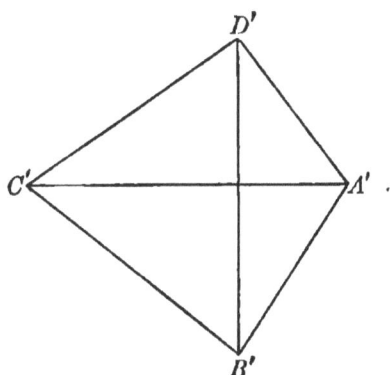

FIG. 47.   FIG. 47 a.

Now the thrusts in the rods AB, BC, CD, DA, and the tensions of the strings AC, BD—that is, the thrusts in the rods $c'd'$, $d'a'$, $a'b'$, $b'c'$, and the tensions of the strings $b'd'$, $a'c'$,—are represented by $C'D'$, $D'A'$, $A'B'$, $B'C'$, $B'D'$, $A'C'$ respectively, and are therefore proportional to CD, DA, AB, BC, BD, AC respectively. Thus the thrusts in the rods and the tensions of the strings are proportional to the opposite sides and diagonals of the quadrilateral.

47. Ex. 4. *Two fine light rods AC, CB rest in a given position, being smoothly jointed to one another at C and to two fixed points at A and B. A given force* *is applied at C in the plane ABC. It is required to determine the stresses in the rods, and to examine the nature of those stresses for different directions of the applied force.*

Let HK represent the applied force. Draw through H straight lines bHb', aHa' in the directions of CB, CA respectively. Through K draw KL parallel to CA to meet bHb' in L. Then KL, LH represent the actions of the rods CA, CB respectively upon C.

If HK is within the angle aHb, both rods are ties; if within the angle aHb', the rod AC is a tie and CB a strut; if within the angle $a'Hb$, the rod AC is a strut and CB a tie; if within the angle $a'Hb'$, both rods are struts.

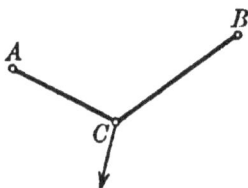 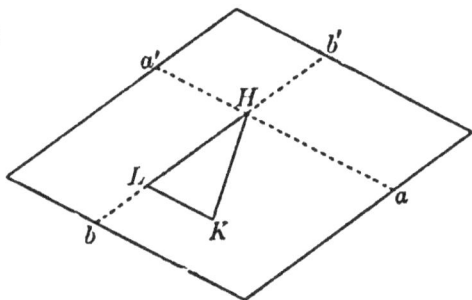

FIG. 48.   FIG. 48 a.

Suppose that the maximum tension and compression that each rod can bear without breaking are known. Take Ha, Hb to represent the maximum tensions of the rods AC, BC respectively, and Ha', Hb' the maximum compressions of the rods AC, BC respectively. Through a, a' draw parallels to bH, and through b, b' parallels to aH. Then the point K must lie within the parallelogram so formed.

EXAMPLES IV.

· 1. A fine light rod AB, of length 15 inches, is capable of turning freely in a vertical plane about the end A, which is fixed. A mass of 60 pounds is suspended from B, and the whole is supported by a horizontal string BC, of length 14 inches, attached to a fixed point C, distant 13 inches from A. Find the tension of the string, the thrust in the rod, and the action at A.

If the string cannot bear a tension greater than 90 pounds' weight, find the least load which, applied at B, will break the string.

Also, if there is no load supported at B, find in what direction a force of 75 pounds' weight must be applied at B, in order to be just on the point of breaking the string.

2. Two fine light rods AC and CB, of lengths 25 feet 8 inches and 17 feet 1 inch respectively, are jointed together at C and to two fixed points A and B, the point B being 9 feet 9 inches vertically above A. A mass of 1 cwt. is suspended from C. Find the stresses in the rods.

If the greatest thrust that the rod AC can bear is the weight of 40 cwt., and the greatest tension that the rod BC can bear the weight of 41 cwt., find the magnitude and direction of the force which, applied at C, will be on the point of breaking both rods simultaneously.

Find also the greatest load which can be sustained at C.

3. Four equal rods, of no appreciable weight, are hinged together to form the rhombus $ABCD$, and the hinges at B and D are joined by another equal rod BD, of no appreciable weight. If the rhombus is supported at A, and a mass of 1 cwt. is suspended from C, find the thrust in BD.

4. If, in the preceding example, the cross rod BD is half as long again as each of the other rods, find the stress in each rod under the same load as before.

5. $ABCD$ is a framework of four light rods loosely jointed together, AB and AD being each of length 4 feet, BC and CD each of length 2 feet. A mass of 100 pounds is attached to the

hinge C, and the whole framework, which is stiffened by a light rod of length 3 feet connecting the hinges B and D, is suspended from A. Find the thrust in the rod BD.

6. $ABCD$ is a framework of four light rods loosely jointed together, AB and AD being each of length 4 feet, BC and CD each of length 2 feet. The hinge C is connected with A by means of a fine string of length 5 feet, and the whole is placed on a smooth horizontal table. If the hinges B and D are pressed towards one another by two forces each equal to 25 pounds' weight in the straight line BD, find the tension of the string.

7. Four fine light rods, of lengths 20, 15, 20, 15 inches, are smoothly hinged together to form a parallelogram $ABCD$, and the hinges B and D are connected by another fine light rod of length 31 inches. If the system is suspended from A, and a mass of 68 pounds is attached at C, find the position of equilibrium and the stress in each rod.

8. In the preceding example the cross rod BD is of length 17 inches. Find the position of equilibrium and the stress in each rod when a mass of 62 pounds is attached at C.

9. A fine light rod HK, of length 15 inches, is connected with a fixed point M by two fine light strings HM and KM, of lengths 13 and 4 inches respectively. Another fine light string, of length 27 inches, has its extremities attached to the points H and K, and supports a mass of 60 pounds at the point N, situated 14 inches along the string from H. The whole is allowed to rest in a vertical plane. Find the position of equilibrium, the tension of each portion of string, and the thrust in the rod.

10. Four fine light rods are smoothly jointed at their extremities to form a parallelogram. The opposite joints are connected by two fine light strings in a state of tension. Prove that the thrusts of the rods and the tensions of the strings are proportional to the lengths of the rods and strings respectively.

11. Four fine light rods are smoothly jointed at their extremities to form a trapezium. The opposite joints are connected by two fine light strings in a state of tension. Prove that the thrusts in the parallel rods are inversely proportional to the lengths of

those rods, and that the thrusts in the non-parallel rods and the tensions of the strings are proportional to the lengths of those rods and strings respectively.

12. Four fine light rods are smoothly jointed at their extremities to form a quadrilateral $ABCD$. The opposite joints are connected by two fine light strings AC, BD in a state of tension. $A\delta$ is drawn parallel to BC to meet BD in δ, and $B\gamma$ is drawn parallel to AD to meet AC in γ. Prove that $\gamma\delta$ is parallel to CD; also that the thrusts in the rods AB, BC, CD, DA and the tensions of the strings AC, BD are proportional to AB, $A\delta$, $\gamma\delta$, $B\gamma$, $A\gamma$, $B\delta$ respectively.

13. Four fine light rods are smoothly jointed at their extremities to form a quadrilateral framework. The opposite joints are connected by two fine light strings in a state of tension. Prove that if the thrusts in two opposite rods are proportional to the lengths of those rods, the other two rods must be parallel.

14. Four fine light rods are smoothly jointed at their extremities to form a quadrilateral $ABCD$. The opposite joints are connected by two fine light strings AC, BD in a state of tension. Prove that if the diagonal BD bisects the diagonal AC, (i.) the thrusts in the rods AB, BC are proportional to the lengths of those rods; and (ii.) the thrusts in the rods CD, DA are proportional to the lengths of those rods.

Conversely, if the thrusts in the rods AB, BC are proportional to the lengths of those rods, prove that (i.) DB bisects AC, and (ii.) the thrusts in the rod CD, DA are proportional to the lengths of those rods.

CHAPTER V.

FINE LIGHT STRINGS IN CONTACT WITH SMOOTH SURFACES.

48. Let a portion of a fine light string, in a state of tension, rest in contact with a smooth surface into which it does not penetrate. This portion takes up the shape of the surface against which it rests, and the surface, being smooth, presses it, at every point where it touches it, in a direction perpendicular to the tangent at that point, that is in the direction of the normal at that point.

It is usual for the beginner to *assume* that under these circumstances the tension of the string is the same at every point. We offer here a proof of this proposition by the graphical method.

49. Let the portion AB of the string $HABK$ rest against a smooth surface. Divide this portion up into a number of parts in the points P_1, P_2, P_3, P_4.

Let T, T_1, T_2, T_3, T_4, T' be the measures of the tensions at the points A, P_1, P_2, P_3, P_4, B respectively.

Draw HAQ_1, $Q_1P_1Q_2$, $Q_2P_2Q_3$, $Q_3P_3Q_4$, $Q_4P_4Q_5$, Q_5BK the tangents at A, P_1, P_2, P_3, P_4, B respectively.

Consider the equilibrium of the portion AP_1 as a rigid body. The external forces acting upon it are,—

the tension T in direction AH, the tension T_1 in direction P_1Q_2, and the forces with which the surface presses it outwards at every point. These pressures together balance the first two forces and must therefore be equivalent to a resultant pressure equal and opposite to the resultant of the two forces T and T_1.

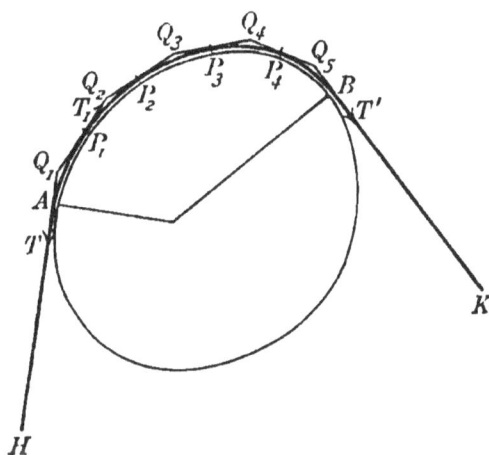 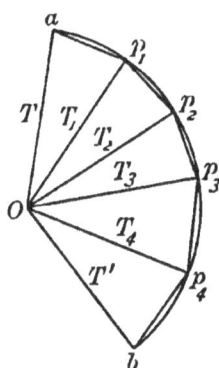

FIG. 49.   FIG. 49 a.

Hence a force diagram aOp_1 can be constructed, in which aO represents T, Op_1 represents T_1, and p_1a represents the resultant pressure of the surface upon the portion of string AP_1.

Similarly, if we consider the equilibrium of the portion P_1P_2, we have the force diagram p_1Op_2, in which Op_2 represents T_2, and p_2p_1 represents the resultant pressure of the surface upon the portion of string P_1P_2.

Proceeding in this way, we have the force diagram indicated above, in which Op_3, Op_4, Ob represent the tensions at P_3, P_4, B respectively; and p_3p_2, p_4p_3, bp_4 represent the resultant pressures of the surface upon the portions of string P_3P_2, P_4P_3, BP_4 respectively.

Now suppose the points P_1, P_2, P_3, ... to be indefinitely increased in number and taken indefinitely close together. Then the straight lines Oa, Op_1, Op_2, Op_3, ... become indefinitely close together, and $ap_1p_2p_3$... becomes ultimately a continuous curve ab.

Now the pressure on any element P of the string is represented by the little element of the curve ba at p. But the pressure is normal at P, and therefore perpendicular to the tension at P, which is represented by Op.

∴ the direction of the curve apb at p is perpendicular to Op.

In other words, Op is the normal at any point p of the curve apb.

∴ the curve apb must be a circle with centre O.

∴ Op is constant for all positions of p and equal to Oa or Ob.

In other words, the tension at any point P of the string is the same as at A or B.

50. *Resultant pressure between a fine light string, in a state of tension, and a smooth peg, round which it passes.*

Let the string ABC, in a state of tension, pass round a smooth peg at B. The tension of the string is the same at every point; let its measure be T.

Consider the equilibrium of the portion of string HBK in the neighbourhood of the peg. The external forces acting upon it are—the tension T at H in the direction HA, the tension T at K in direction KC, and the forces with which the peg presses the string at every point of contact. These pressures must produce a resultant pressure equal and opposite to the resultant of the first two forces.

Hence a triangle of forces LMN can be constructed, in which LM and MN are each of length T units, and parallel to HA and KC respectively, and in which NL represents the resultant pressure of the peg upon the string. This is equally inclined to LM and MN, and is therefore in the opposite direction to the internal

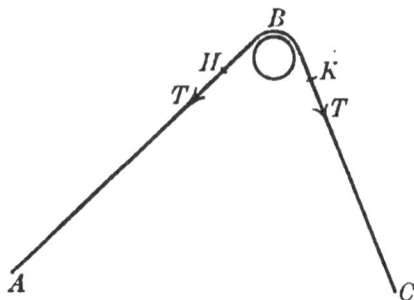

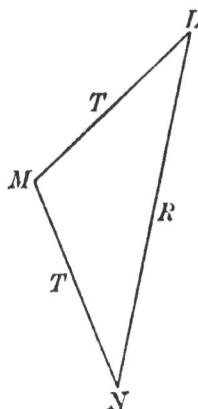

FIG. 50.   FIG. 50 a.

bisector of the angle ABC. Also the line of action of this pressure passes through the point of intersection of AH and CK.

The pressure of the string upon the peg is equal and opposite to the pressure of the peg upon the string, and is therefore in the direction of the bisector of the angle ABC.

51. Ex. 1. *A fine light string has its extremities attached to two masses, each weighing 5 pounds, and passes over two small smooth pegs H, K. The peg K is situated 7 inches farther from the ground than H, and 24 inches horizontally to the right of H. Find the resultant pressures of the string upon the pegs.*

In the position of equilibrium, the portion of the string between H and K is straight, and the other two portions hang vertically downwards. Also, the tension of the string is everywhere 5 pounds' weight.

Having constructed the space diagram to scale, we mark the three portions of the string with the letters oa, ob, oc, as in the figure. Draw the straight line ab bisecting the angle between the portions of the string·

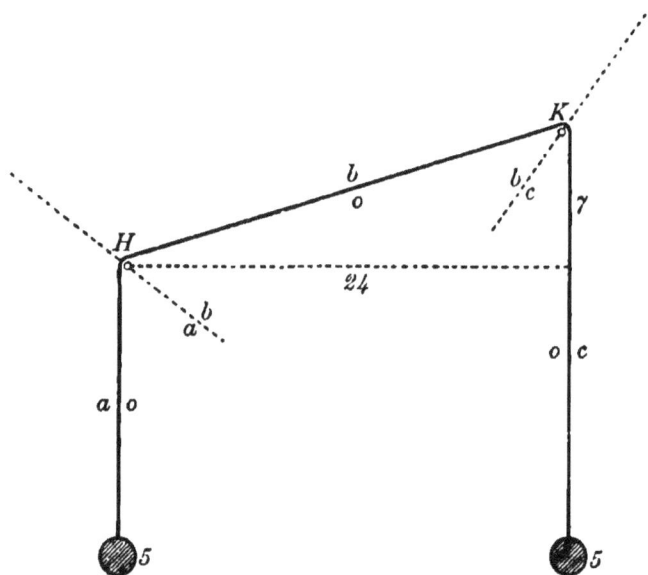

FIG. 51.

at H, and the straight line bc bisecting the angle between the portions of the string at K.

Draw OA, of length 5 units, vertically downwards, to represent the tension of the string oa. Through O and A draw straight lines OB, AB parallel to ob, ab respectively. This gives the point B. Draw BC parallel to bc to meet AO produced in C. Then $OABO$ (this way round) and $OBCO$ (this way round) are the triangles of forces for the portions of string in the neighbourhood of H and K respectively.

On measurement we find that AB and BC are of length 6 and 8 units respectively. Thus, the pressure

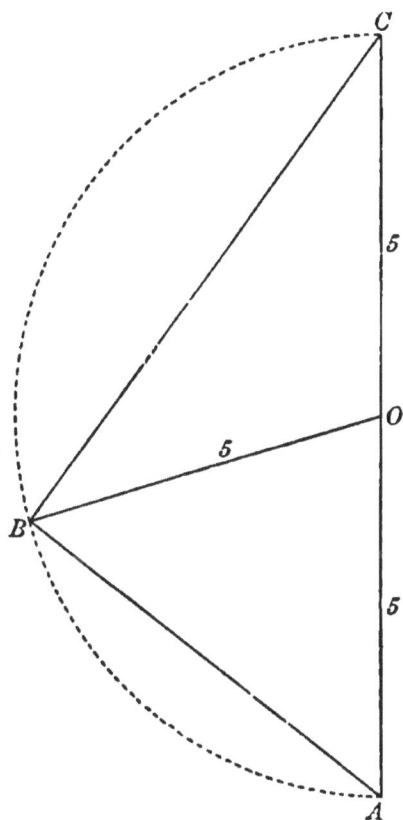

FIG. 51 a.

between the string and the peg H is 6 pounds' weight, and the pressure between the string and the peg K is 8 pounds' weight.

Without drawing the lines ab and bc, we can construct the force diagram by describing a circle with centre O

and radius OA, and by drawing OB in the direction of
KH to meet this circle in B.

52. Ex. 2. *A fine light string, 31 inches long, passes
through a small ring of mass 4 ounces, and has its
extremities fixed at two points 25 inches apart in the
same horizontal line. Find the magnitude of the
horizontal force which, applied to the ring, will cause
it to rest at a point 7 inches from the nearer end of
the string. Also determine the tension of the string.*

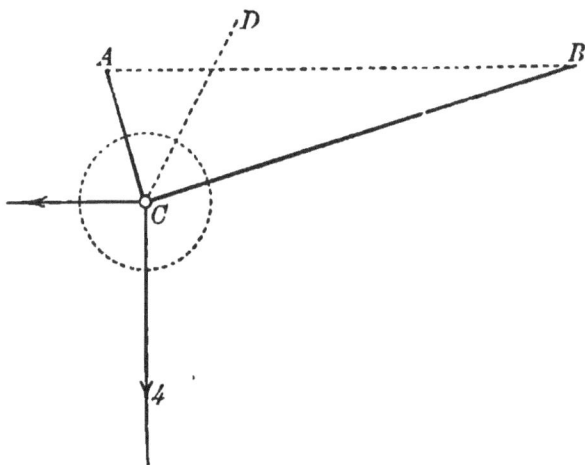

FIG. 52.

Take A and B to represent the two given points 25
inches apart. In the position of equilibrium the two
portions of the string are straight, and, as the string
rests against the smooth surface of the ring, its tension
is the same on either side of the ring, and therefore
the same throughout. The position of the ring will be
at C, which is 7 inches from A and 24 inches from B.
Having found the position of C, draw CD bisecting
the angle BCA.

Consider the equilibrium of the portion of matter as one rigid body contained within a closed curve drawn round C. The external forces acting on this portion of the system are—its weight, which amounts in all to 4 ounces' weight, as the string itself is of no appreciable weight, the equal tensions of the strings along CB and CA, and the unknown horizontal force. The equal

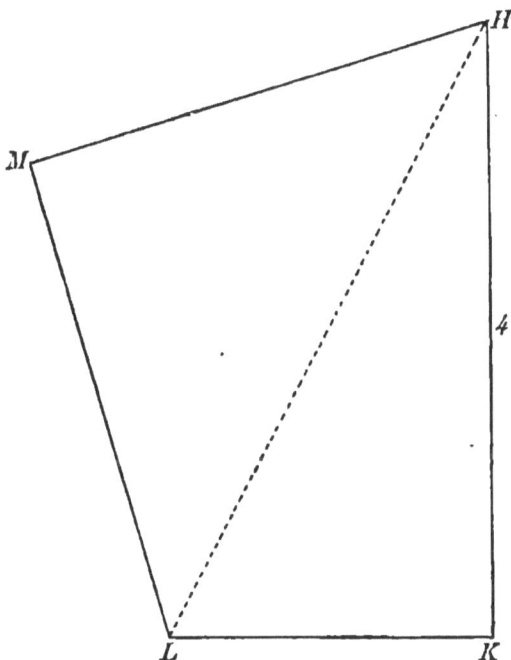

FIG. 52 a.

tensions of the strings can be replaced by a single force of unknown magnitude acting along CD. Thus we have reduced the forces acting upon the portion of the system under consideration to three forces acting along known lines.

Draw HK vertically downwards, of length 4 units, to represent the weight of the ring. Through H draw a straight line parallel to DC to meet the horizontal through K in L. Through H and L draw straight lines parallel to BC and CA respectively, to meet in M. Then KL represents the horizontal force, and LM, MH each represent the tension of the string.

On measuring the lines of the force diagram, we find that the horizontal force is 2·19 ounces' weight, and the tension of the string is 3·23 ounces' weight.

53. Ex. 3. *A fine straight rod HK, of no appreciable weight and of given length, has two pieces of fine light string of given lengths attached to it in the manner indicated in the diagram. One of these strings passes through a small smooth ring L of no appreciable weight, which is connected by means of another fine light string*

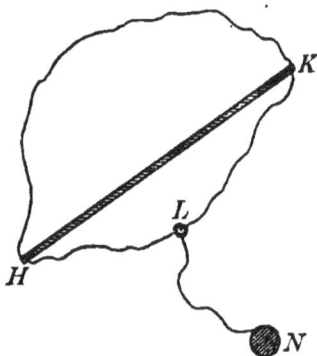

FIG. 53.

to a mass N of given weight. The other string is then placed over a small smooth fixed peg M. Prove that, in a position of equilibrium, the rod is either vertical or horizontal, and show how to determine in each case the tension of the string and the thrust in the rod.

In a position of equilibrium both strings are in a state of tension, and hence the portions HL, LK, KM, MH

are all straight; also, the string supporting the mass N
hangs vertically below L.

As we are unable at the outset to draw accurately
a space diagram for the system in equilibrium, we
assume a position of equili-
brium, making no attempt to
construct the diagram to scale,
and endeavour to ascertain the
properties of the figure.

We will consider first the
equilibrium of the rod HK, the
strings HMK, HLK, and a portion LX of the string LN, all
together as one rigid body. The
only forces acting externally
upon this portion of the system
are the pressure of the peg
upon the string at M and the
tension of the string LX at X. These two forces must
be equal and in opposite directions along the same
straight line.

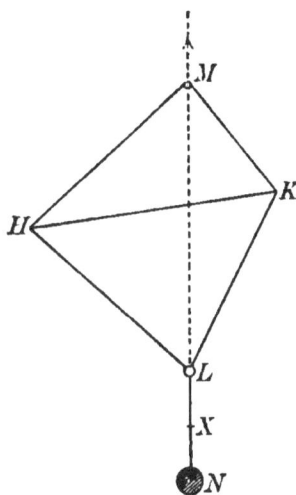

∴ ML is vertical and in the same straight line with LN.

As the strings HMK and HLK rest against smooth
surfaces at M and L respectively, their tensions are
in each case the same throughout. Now consider the
equilibrium of a portion of the system included within
a closed curve drawn round L. This shows at once
that the vertical through L bisects the angle HLK.
Similarly, considering the equilibrium of a portion of
the string in the neighbourhood of M, we see that the
pressure of the peg upon the string at M, already shown
to be vertical, must balance the two equal tensions of

FIG. 54.

D.S.   F

the string, and therefore the vertical through M bisects the angle HMK.

Thus, the position of equilibrium is such that LM is vertical, and bisects the angles between the strings at L and M respectively.

I. The strings HM, MK may both be vertical. The points H and K are then both in the line ML, and the strings HL, LK are also vertical.

This gives two positions of equilibrium, namely, with H above K, or with K above H.

The diagram in each case is readily drawn as the lengths of the strings HMK, HLK and of the rod are all known.

II. If HM, MK are not vertical, they are equally inclined to ML, and the points H, K are on opposite sides

FIG. 55.   FIG. 56. of ML. In this case, in the triangles HML, KML, the side ML is common, and the angles HML, HLM are respectively equal to the angles KML, KLM.

∴ the triangles HML, KML are equal in all respects.

This shows that the figure is symmetrical with respect to the vertical LM, and hence that HK is horizontal.

The space diagram is now readily constructed to scale; for the lines HM, MK are each half of the given length of the string HMK, and the lines HL, LK are each half of the given length of the string HLK.

Having constructed the space diagram to scale and marked it with the letters a, b, o_1, o_2, as indicated, we draw AB to represent the given weight of the mass N.

AO_1, BO_1, AO_2, BO_2 are then drawn parallel to the lines ao_1, bo_1, ao_2, bo_2 respectively, and joining the points O_1, O_2 thus obtained, we have the complete force diagram.

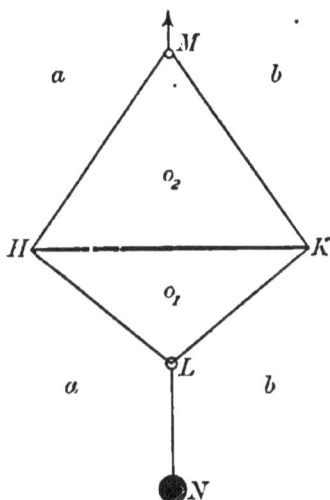   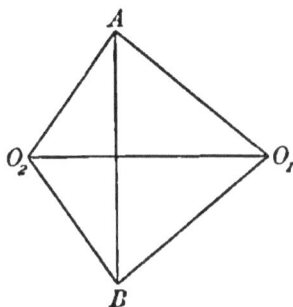

FIG. 57.   FIG 57 a.

In the cases I., where the strings are all vertical, the points O_1, O_2 coincide with the middle point of AB. Therefore the thrust in the rod is zero, and the tension of each string is half the weight of the mass.

54. Ex. 4. *A fine light string, of given length, is passed through a smooth ring, of no appreciable weight or size, and is attached at its extremities to two given points. A force, given in magnitude and direction, is applied to the ring. It is required to find the position of equilibrium and the tension of the string.*

Here, at the outset, we are unable to construct the space diagram, as we do not know the position of equilibrium. We therefore *assume* a position of equilibrium and represent it in a diagram without any attempt at first to construct it to scale.

Let AMB represent the string in the position of equilibrium, A and B being the two fixed points and M the position of the ring.

Let P be the measure of the force applied at M in direction ML, and let CD be drawn P units of length in the given direction of this force; thus ML and CD are parallel.

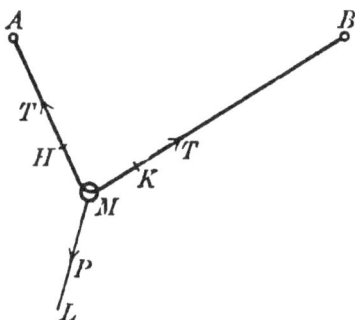
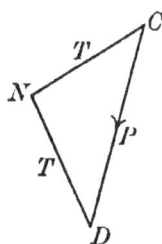

FIG. 58.   FIG. 58 a.

As the string, in passing through the ring, rests against a smooth surface, its tension is the same on one side of the ring as on the other, and therefore its tension is the same throughout. Let T be the measure of this tension.

Consider the equilibrium of the ring, together with the portion HMK of the string in its immediate neighbourhood, as one rigid body. The forces acting externally upon this are,—the tension T at H in direction HA, the tension T at K in direction KB, and the force P. Hence our force diagram for this system will be a triangle CDN in which DN, NC are parallel to MA, MB respectively, and each T units of length. Hence CD must be equally inclined to NC and DN, and therefore LM produced must bisect the angle AMB. This gives the following construction for finding the position of M:

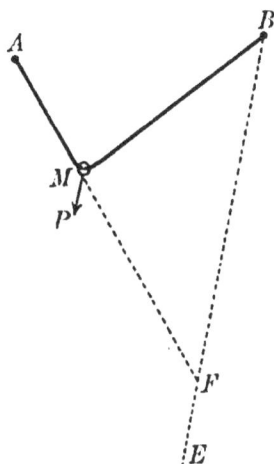

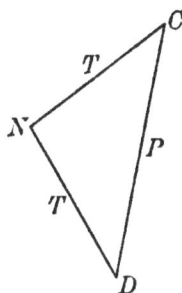

FIG. 59.   FIG. 59 a.

Through B draw BE in the direction of CD. With centre A, and radius representing the given length of the string, describe a circle cutting BE in F. Let the straight line which bisects BF at right angles meet AF in M. Then M is the position of the ring.

The student will have no difficulty in proving that AMB is of the proper length, and that the straight line drawn through M in the direction of DC is the bisector of the angle AMB.

Draw CN parallel to BM and DN parallel to MA, and let CN and DN meet at N. Then, measuring either of the two lines DN and CN, we have the tension of the string.

We must see that the point F is taken in BE, and not EB produced. Also the string must evidently be longer than AB; this being so, we get one, and only one, position of equilibrium.

55. Ex. 5. *To the extremities of a fine light string, which passes round two small smooth pegs given in position, are applied two given forces in given directions, one at each end. To a point of the string between the .pegs is applied a third given force in a given direction. It is required to find the position of equilibrium and the pressures between the string and the pegs.*

Without attempting at the outset to construct a space diagram to scale, suppose the string takes up the position *HALBK*, *A* and *B* being the two given pegs.

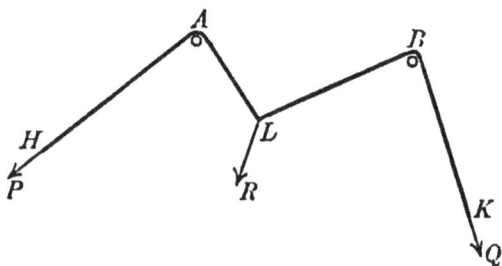

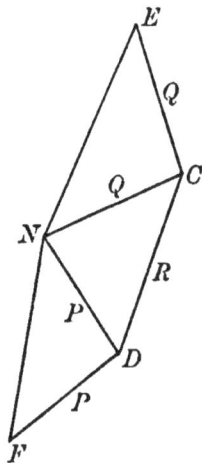

Fig. 60.   Fig. 60 a.

Let *P* and *Q* be the measures of the given forces applied at *H* and *K* respectively, and *R* the measure of the given force applied at *L*. The directions of *AH* and *BK* are of course the given directions of *P* and *Q*, but the directions of *AL* and *BL* are at present unknown.

Since the peg *A* is smooth, the tension of the string *HAL* is the same at every point, and therefore its measure is P. Similarly the measure of the tension of the string LBK is Q at every point.

Hence the force diagram for the point L will be a triangle CDN in which CD is R units of length in the given direction of the force R; and DN, NC are respectively P, Q units of length and parallel to LA, LB respectively.

Also, if CE be drawn in the direction of KB and equal to CN, and if DF be drawn in the direction of AH and equal to DN, then NF and EN will represent the pressures of the string upon the pegs at A and B respectively.

Hence we have the following construction: Draw EC, CD, DF parallel to the directions of the given forces Q, R, P and of lengths Q units, R units, P units respectively. With centre C and radius CE describe a circle, and with centre D and radius DF describe another circle intersecting the first circle in N.

Draw AL and BL parallel to ND and CN respectively. This gives the position of L, and the directions of AH, BK are already known.

Then, measuring NF and EN, we have the pressures at A and B respectively.

As the circles may intersect in two points, this apparently gives two solutions, but the student will readily see that if he takes the point of intersection of the circles on the other side of CD he will get an inadmissible result.

88 FINE LIGHT STRINGS.

EXAMPLES V.

1. One end B of a fine light cord is fixed; the cord passes over a small smooth fixed peg A in the horizontal line through B, and supports at its other end C a mass of weight P; find the magnitude and direction of the pressure on the peg.

2. A fine light string ACB, of length 14 inches, has its extremities attached to two fixed points A and B, situated 10 inches apart on a smooth horizontal table. To a point C of the string, 8 inches from A, is knotted another fine light string CD, which passes over the smooth edge of the table and supports at its free end a mass of 20 pounds. Find the tensions in AC and BC, supposing that DC produced passes through the middle point of AB.

3. An endless fine light string, of length 3 feet, on which a small heavy ring of weight W is capable of sliding freely, is supported on two small fixed pegs situated 1 foot apart in a horizontal line. Find the pressure between the string and each peg.

4. A fine light string $ABCD$ has one extremity A fixed, and passes over two small smooth pegs at B and C, supporting at its free end D a mass of 10 pounds. If ABC is an equilateral triangle, having the side AC vertical and A uppermost, determine the pressure between the string and each peg.

5. A fine light string has one end attached to a fixed point A. It passes over a small smooth peg B, situated 1 foot 4 inches to the right and 1 foot above A, and supports at its other extremity a mass of 25 pounds. Find the pressure between the string and the peg.

6. A fine light string has one extremity attached to a fixed point A, passes over a small smooth peg B, and supports at its other extremity a mass of 20 pounds. The peg B is situated 5 inches to the left of A and 12 inches above it. Find the pressure on the peg.

7. B and C are two smooth rings fixed in space at a distance apart equal to 13 inches, B being 10 inches and C 15 inches above the ground. A fine light string $ABCD$ passes through the rings, and supports at its extremities masses weighing 10 pounds each. Find the pressures between the string and the rings.

8. A fine light string, of length 64 inches, passes over two small smooth pegs fixed 30 inches apart in a horizontal line. Both extremities of the string are attached at the same point to a mass of 20 pounds. Find the pressure between each peg and the string.

9. A fine light string, 28 inches long, passes through a small smooth ring, to which is attached a mass of 24 pounds, and has its extremities fixed at two points situated 14 inches apart in the same horizontal line. Find the magnitude of the horizontal force which, applied to the ring, will cause it to rest at a point 13 inches from the nearer end of the string. Also determine the tension of the string.

10. A fine light string, of length 25 inches, has its extremities attached to two points, situated in a horizontal line 7 inches apart. A small smooth ring, of mass 3 pounds, is capable of sliding freely on the string. Find the tension of the string in the position of equilibrium.

11. A fine light string, 31 inches long, passes through a small ring of 4 ounces' weight, and has its extremities fixed at two points 25 inches apart in the same horizontal line. Find the tension of the string in the position of equilibrium.

12. A fine light string, of length 15 inches, is passed through a small smooth ring C of no appreciable weight, and is attached at its extremities to two fixed points A and B. The point B is situated 3 inches farther from the ground than A and 1 foot horizontally to the right of A. If a mass of 12 pounds is connected with the ring C by means of another fine light string, find the position of equilibrium and the tension of the string.

13. A fine light string, of length 32 inches, is passed through a small smooth ring of no appreciable weight, and is attached at its extremities to two points A and B situated 24 inches apart. A force of 16 pounds' weight is applied to the ring in a direction making an angle of 53° with BA. Find the position of equilibrium and the tension of the string.

14. A fine endless string, of length 20 inches, rests on three smooth pegs A, B, C, the pegs B and C being situated in a horizontal line 6 inches apart, and A 4 inches vertically over the middle point

of *BC*. To a point *D* of the loop of string below *BC* is attached a mass of 8 pounds. Find the pressures on the pegs and the tension of the string.

15. *A*, *B*, *C* are three smooth pegs fixed in a vertical plane, *A* being 3 feet vertically above the middle point of *BC*, which is horizontal and 8 feet long. A string, 20 feet long, passes round the three pegs, and has its extremities attached at the same point to a mass of 12 pounds. Find the tension of the string and the resultant pressures on the pegs.

16. *A*, *B*, *C* are three smooth pegs fixed in a vertical plane, *A* being 3 feet vertically below *B* and 4 feet horizontally to the right of *C*. A fine light string, 13 feet long, passes round the three pegs, and has its extremities attached at the same point to a mass of 12 pounds. Find the tension of the string and the resultant pressure on each peg.

17. A fine light string *OABO*, 2 feet long, passes round two smooth small pegs at *A* and *B*, situated 8 inches apart on a smooth horizontal table. The two ends of the string are knotted together at *O*, 6 inches from *A*. In what direction must a horizontal force of 16 pounds' weight be applied to the knot *O*, in order that the string may remain stretched without slipping over the pegs? If the force has this direction, find the tension of the string and the pressures upon the pegs.

18. A fine light string *ACB*, of length 20 inches, has its extremities attached to two points *A* and *B*, situated 16 inches apart in a horizontal line. To the middle point *C* of the string is attached a small smooth ring of no appreciable weight. Another string has one extremity attached at *D*, 21 inches vertically below *B*, and passes through the ring, supporting at its other extremity a mass of 51 pounds. Find the tensions of *AC* and *BC*.

19. A fine light string, passing over two smooth parallel bars one foot apart in a horizontal plane, has two masses each weighing 25 pounds fastened to its extremities, and another mass weighing 14 pounds is attached to a point *P* of the string between the bars; in the position of equilibrium find the depth of *P* below the level of the bars; find also the magnitude of the pressure upon each bar.

20. A fine light string $ABCD$ is attached at one extremity to a fixed point A. It passes through a small smooth ring B, of mass 1 pound, and over a smooth peg C, and supports at its extremity D a mass of 4 pounds. Find the pressure between the string and the peg in the position of equilibrium.

21. A fine straight rod AB, of no appreciable weight and of length 30 inches, has two pieces of fine string ACB and ADB, of lengths 50 and 34 inches respectively, attached to it at A and B. The shorter string passes through a small smooth ring, of no appreciable weight, which is connected by another fine string with a mass of 16 pounds. The longer string is placed over a small smooth fixed peg. Show that, in the position of equilibrium, the rod is either vertical or horizontal, and in each case determine the tensions of the strings.

22. A, B, C are three points in a vertical plane, A and C lying on opposite sides of the vertical line through the highest point B. A fine light string $ADBCDE$, having one end fixed at A, passes in succession through a light smooth ring D, round pegs at B and C, again through the ring, and is attached to a heavy mass at its free extremity E. Prove that, in the position of equilibrium, the ring and the mass hang vertically below B.

23. A fine light string $AXBC$ is attached at one extremity to a fixed point A. It passes over a smooth peg B, and supports at its extremity C a mass of given weight. Show how to determine at what point X, between A and B, another mass of given weight must be attached, in order that, in the position of equilibrium, AX may be in a given direction. Show that there may be two solutions, but that the length of XB is the same in both cases.

CHAPTER VI.

EQUILIBRIUM OF A PARTICLE RESTING IN CONTACT WITH A SMOOTH SURFACE OR CURVE.

56. Suppose that a particle, acted upon by a system of forces, rests in contact with a smooth surface at P. Then, in addition to the other forces that act upon it,

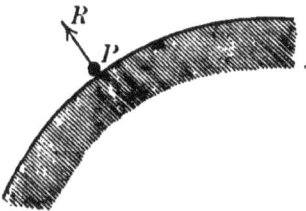

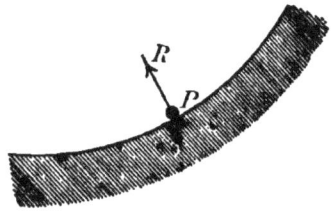

FIG. 61.   FIG. 62.

there is the force R, with which the surface resists any tendency that the particle may have to penetrate it. If the surface is smooth it cannot resist any tendency to slide over it; it can only press outwards in the direction of the normal at P. If we assume that the material of which the surface is composed is sufficiently strong for all purposes, then there is no limit to the magnitude of this force R. This force, which is called the reaction of the surface, is a self-adjusting force, that is, it will be of the magnitude required to preserve equilibrium, if equilibrium is possible. Hence, for equilibrium, it is necessary and sufficient that the resultant of all the other forces acting upon the particle should be in the direction of the normal at P, and inwards—that is, towards the surface.

If, instead of the surface, we have a material plane curve on which the particle can slide, as a bead threaded on a wire, or a fine tube in which the particle is placed; and if the forces acting on the particle are all in the plane of the curve, then the reaction of the curve may be in either direction, inwards or outwards, and for equilibrium it is only necessary that the resultant of all the other forces acting upon the particle should be perpendicular to the tangent at P.

If we suppose applied to the particle a force R identical with the reaction of the surface or curve, we may suppose the latter removed altogether, and then consider the particle as in equilibrium under the influence of the given system of forces that act upon it together with the force R. Thus the consideration of the equilibrium of the particle is the same as for a particle free to move, with this difference,—that there are certain limitations upon the direction of the force R.

57. Equilibrium of a Heavy Particle on a Smooth Inclined Plane.

A particle of given weight is placed on a smooth plane inclined at a given angle to the horizon, and is sustained by some force applied to it in some direction which is in a vertical plane with the line of greatest slope. It is required to represent graphically the different values of the sustaining force corresponding to the different directions in which it may be applied, and to find in each case the pressure between the particle and the plane.

Let O be the position of the particle, AOB the line of greatest slope making the given angle with the horizontal AC. Suppose the particle is sustained by a force whose measure is P applied in direction OH. Let W be the measure of the weight of the particle, and let it be represented by EF drawn vertically downwards W units of length.

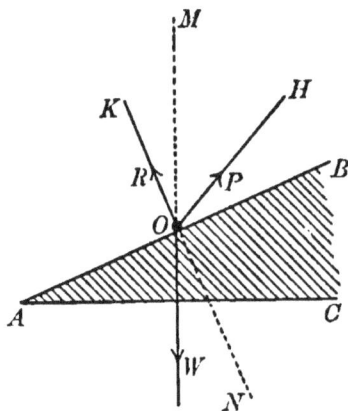

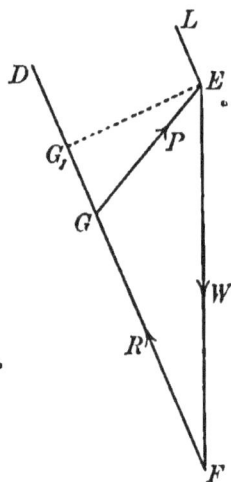

FIG. 63.   FIG. 63a.

The only effect of the presence of the inclined plane is to produce a normal reaction outwards, of such magnitude as to balance, if possible, the other two forces acting on the particle. Let R be the measure of the reaction of the plane, and let OK be the straight line drawn from O away from the plane, and in a direction perpendicular to it. Then the two forces P and W are in equilibrium with a force R, which may be of any magnitude, but must be in the direction OK.

Draw FD a straight line of unlimited length from F in the direction OK. The triangle of forces for the particle at O will be a triangle FGE, in which G is some point in FD, and GE is parallel to OH, and of length P units.

Draw EL in the direction FD. Then we see that, as G may have any position in FD, GE may have any direction between FE and LE. Hence, drawing OM vertically upwards, and producing KO to N, we see that the sustaining force may be applied in any direction between OM and ON. The smallest value of P is obtained by drawing EG_1 perpendicular to FD. Then G_1E is parallel to AB. Hence, if the sustaining force is to be as small as possible, it must be applied straight up the plane, and its measure is P_1, where G_1E is P_1 units of length.

If the direction of the sustaining force be some given straight line OH between OM and ON, we have merely to draw EG parallel to HO to meet FD in G. Then, measuring FG and GE, we have the measures of the reaction of the plane, and of the sustaining force respectively.

If the measure of the sustaining force be some given number P greater than P_1, we describe a circle with centre E and radius P_1 units of length, and this will cut FD in two points, giving two directions for the sustaining force equally inclined to OB and on opposite sides of it.

We see from the force diagram, that if the sustaining force be applied in any direction between OB and OM, then its value will be less than that of the weight of the particle.

58. Ex. 1. *A fine straight smooth rod AB, of length 2 feet 6 inches, is fixed with the end A 1 foot 6 inches farther from the ground than the end B. A small smooth ring P, of no appreciable weight, is capable of sliding on the rod, and is connected, by a fine light string 17 inches long, with a point C fixed 10 inches vertically below A. Another fine light string has one end attached to the ring P, and the other end to a mass D of 100 pounds. Find the tension of the string PC, and the pressure between the ring and the rod in the position of equilibrium.*

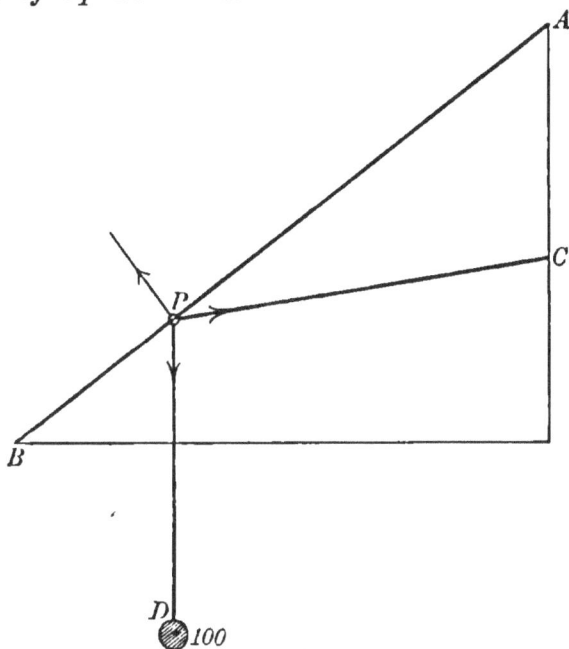

FIG. 64.

In the position of equilibrium, the two strings are straight, and the string PD is vertically downwards.

Having constructed the space diagram to scale, draw HK vertically downwards of length 100 units, to represent the tension of the string *PD*. Through *K* draw *KL* perpendicular to *AB*, to meet the straight line drawn through *H* parallel to *CP* in *L*. Then *HKLH* (this way round) is the triangle of forces for the ring.

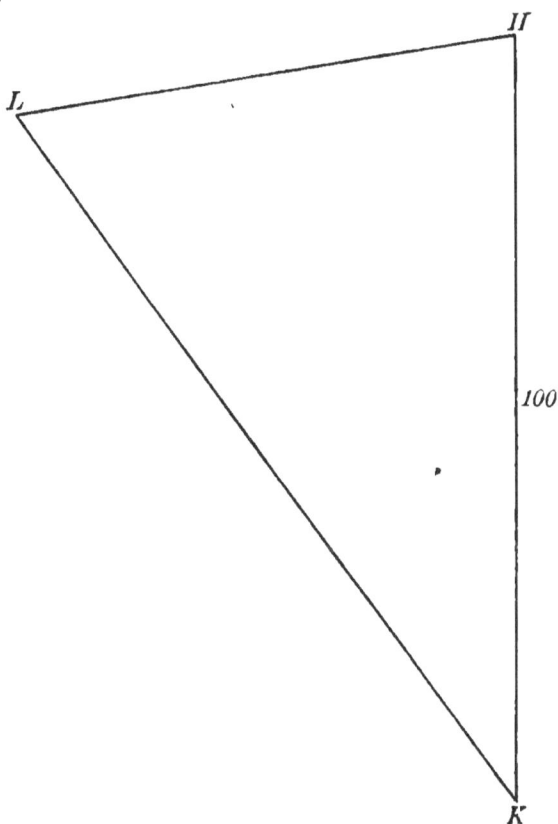

FIG. 64 a.

On measuring, we find that *KL* and *LH* are of lengths 112 and 68 units respectively. Therefore the pressure between the ring and the rod is equal to the weight of 112 pounds, and the tension of the string *PC* to the weight of 68 pounds.

D.S.   G

59. Ex. 2. *A fine straight smooth rod AB is fixed in a given position inclined to the vertical with the end A uppermost. A small smooth ring P, of no appreciable weight, is capable of sliding on the rod; and a fine light string having one end fixed at the point C, situated at a given distance vertically below A, passes through the ring and supports at its other extremity D a mass of given weight. Prove that in the position of equilibrium AC = CP, and show how to find the pressure between the ring and the rod.*

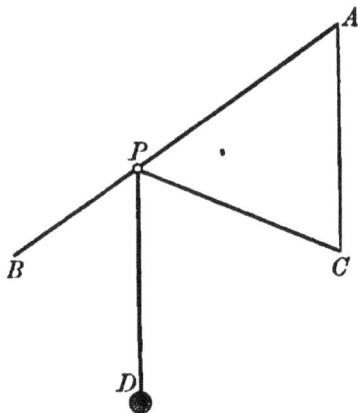

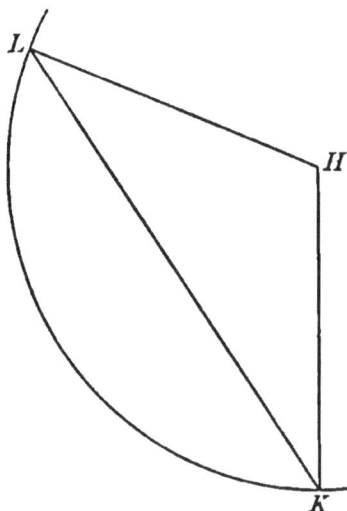

FIG. 65.   FIG. 65 a.

In the position of equilibrium, the tension of the string is at every point equal to the given weight of the mass, the two portions are straight, and the part PD is vertically downwards; but at present we do not know the position of P.

Draw HK vertically downwards of such length as to represent the weight of the mass. With centre H and radius HK, describe a circle meeting the straight line drawn through K perpendicular to AB in L.

Then $HKLH$ (this way round) is the triangle of forces for the system consisting of the ring and of the portion of string in its immediate neighbourhood.

Hence, to find the position of P, we draw CP parallel to HL to meet AB in P.

As HK, HL are equally inclined to the straight line LK which is perpendicular to AB, they are also equally inclined to AB.

$\therefore AC, CP$, which are respectively parallel to HK, HL, are also equally inclined to AB.

$$\therefore AC = CP.$$

Also, measuring KL, we have the pressure between the rod and the ring.

60. Ex. 3. *AB is a smooth straight wire fixed in a given inclined position with B uppermost. A small heavy ring of given weight, capable of sliding freely on the wire, is connected with B by a fine string of given length, which passes through a second small smooth ring of given weight, hanging freely on the string. It is required to find the tension of the string, and the pressure between the ring and the wire.*

Assume a position of equilibrium BCD. Let W_1, W_2 be the measures of the weights of the rings, and R the reaction of the wire, which is perpendicular to AB. The tension of the string is the same throughout; let its measure be T.

Take EF, FG vertically downwards, of length W_1, W_2 units respectively. Then, if EFK be the triangle of forces for the point C, K will lie on the straight line which bisects EF at right angles. Also KFG will be the triangle of forces for the point D, so that GK is perpendicular to AB.

Hence, having taken the points E, F, G, we proceed to complete the force diagram by drawing GH perpendicular to BA, meeting in K the line bisecting EF at right angles. Then, measuring KE and GK, we have the tension of the string and the reaction of the wire.

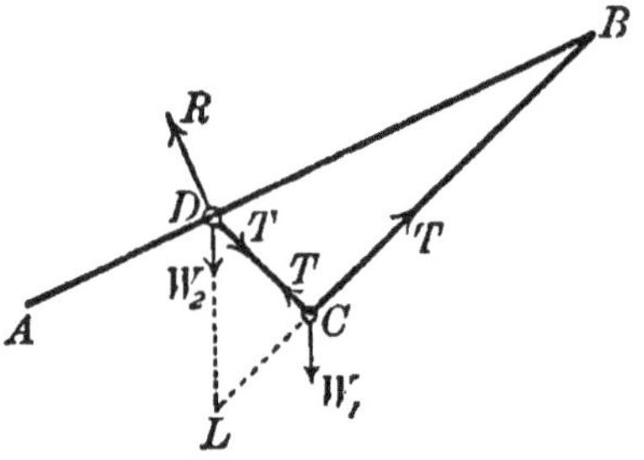

FIG. 66. 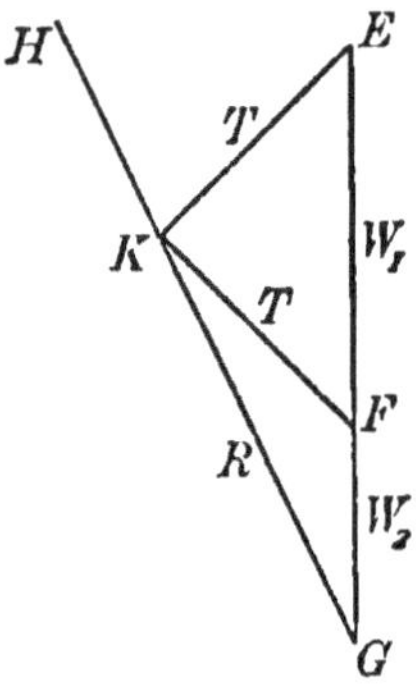   FIG. 66 a.

To finish the space diagram, we draw BL parallel to EK, and of such length as to represent the given length of the string; then we draw LD parallel to GF to meet AB in D, and DC parallel to KF to meet LB in C. Thus we have the positions of the rings and of the string.

61. Ex. 4. *A smooth circular hoop is fixed in a vertical plane. Two small smooth rings of given weight, each capable of sliding freely on the hoop, are connected by a fine string of given length less than the diameter of the hoop. It is required to find the position of equilibrium in which the string is tight, the tension of the string, and the pressures between the rings and the hoop.*

Suppose P and Q are the positions of the rings in equilibrium, O the centre of the hoop. Take AB, BC to represent W_2, W_1, the weights of the rings Q, P, respectively. Let R_1, R_2 be the reactions of the hoop at P and Q respectively. These will be in directions OP, OQ respectively, and therefore equally inclined to the string.

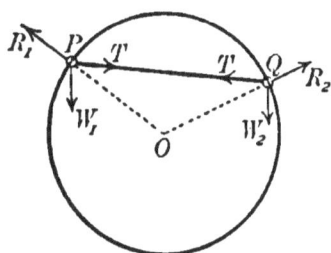

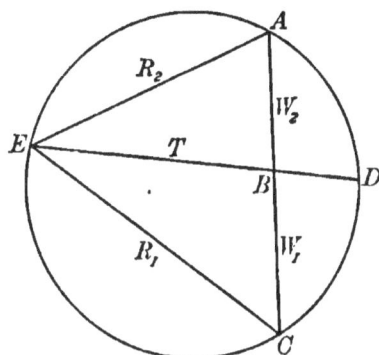

FIG. 67.   FIG. 67 a.

The force diagram for the point Q will be a triangle $ABEA$ (this way round), in which BE is parallel to QP and represents T the tension of the string, and EA is parallel to OQ and represents R_2.

Hence, considering the equilibrium of the ring P, we see that CE must be parallel to OP and represent R_1. Hence EB bisects the angle AEC.

Now the magnitude of the angle POQ can be found, as it is subtended by a chord of given length at the centre of a given circle. Also the angle CEA is the supplement of this angle. Hence we have the following construction for the force diagram: Having drawn AB, BC, describe through A, C the circle $ADCE$ such that the segment ADC contains an angle equal to the angle subtended by the string at the centre O of the hoop. Find D the middle point of the arc ADC, and let DB meet the circumference again in E. To construct the space diagram, we then draw OP and OQ parallel to CE and EA respectively; then we have the positions of the rings P and Q in equilibrium.

EXAMPLES VI.

1. AB is a fixed smooth vertical rod on which a small smooth ring, of mass 3 pounds, is capable of sliding. The ring is supported by a fine light string, of length 10 inches, attached to a fixed point O. If O is at a distance of 8 inches from the rod AB, find the tension of the string and the pressure between the rod and the ring.

2. A is the lowest point of a smooth circular wire fixed in a vertical plane. A small smooth bead, of mass 10 ounces, rests on the wire at P, being supported by a horizontal force F. Find the magnitude of F and the pressure of the wire, if the arc AP subtends an angle of 60° at the centre.

3. A small heavy bead, of mass 20 ounces, is capable of sliding freely on a smooth fixed vertical circular hoop, of radius 5 inches. It is supported by a fine light string, of length 9 inches, attaching it to the highest point of the hoop. Find the tension of the string and the pressure between the bead and the hoop.

4. ABC is a smooth wire fixed with BA vertically upwards, the portions AB, BC being straight, and inclined at an angle of 120°. A small smooth ring, of mass 2 pounds, rests upon the wire at P in BC, where $BP=BA$, and is kept from falling by a fine string connecting it with the point A. Find the tension of the string and the pressure between the ring and the wire.

5. A small ring C rests upon a fixed smooth horizontal rod AB, whose length is 13 feet. To the ring are attached two strings, one of which is 7 feet long and has its other extremity fixed at A, while the other passes over a smooth hook, situated 8 feet

below B, and supports a mass of 20 pounds. Find the tension of the string AC.

6. A small ring of weight W, which can move without friction on a circular wire fixed in a vertical plane, is in equilibrium at a point P, on the lower half of the wire, under the action of a force R in the direction of the tangent at P to the wire. If the pressure of the ring on the wire is equal to $\frac{1}{2}W$, find the magnitude and direction of the force R.

7. A small heavy ring of mass 10 pounds, which can slide freely upon a smooth thin rod AB, is attached to the end A of the rod by a fine string. If the rod is held, with A uppermost, in a position inclined at an angle of 40° to the vertical, find the tension of the string and the pressure between the rod and the ring.

8. Find the force necessary to sustain a particle of mass 5 pounds placed on a smooth plane inclined at an angle of 20° to the horizontal—

(a) when the force is horizontal;

(b) when it acts along the inclined plane.

9. Find the greatest vertical height through which a force 2 pounds' weight can raise a particle of mass 6 pounds by drawing it up a smooth sloping plank 20 feet in length.

10. Find what force, acting horizontally, will support a mass of 30 pounds resting on a smooth inclined plane, the base of the plane being three times its height; also find the pressure on the plane.

11. A small smooth ring P rests upon a fixed smooth horizontal rod AB of length 14 feet. To the ring are attached two strings, one of which is 10 feet long and has its other extremity fixed at C, situated 8 feet vertically below A; the other passes over a small smooth hook, situated 6 feet vertically below B, and supports a mass of 30 pounds. Find the tension of the string PC.

12. A and B are the highest and lowest points respectively of a smooth thin circular wire, of radius 5 inches, fixed in a vertical plane. A small bead P, weighing 5 ounces, is threaded

on the wire, and is attached to A by a fine string of length 8 inches. A second string is attached to the bead and passes over a small smooth peg at B, supporting at its other extremity another bead weighing 3 ounces. Find the tension of AP.

13. A smooth straight rod AB is fixed in a position inclined at an angle of 60° with the vertical, the end B being uppermost. A small smooth ring C, of mass 2 pounds, is threaded upon the wire, and is connected with B by a fine string; a second string is attached to the ring, and passes over a small smooth peg at D fixed vertically below A, supporting at its free end a mass of 10 pounds. If $AC=AD$, find the tension of the string BC and the pressure between the ring and the wire.

14. A smooth straight rod AB is fixed in a position making an angle of 60° with the vertical. A fine light string, one end of which is attached to A, the highest point of the rod, passes through a small smooth ring D, to which a mass of 30 pounds is attached, and the other end of the string is attached to a small smooth ring C capable of sliding freely along the rod. Find the angle ADC and the tension of the string in the position of equilibrium, the weights of the rings C and D being inappreciable. Find also the distance of C from A in the position of equilibrium, given that the length of the string is 6 inches.

15. ACB is a smooth thin wire in the form of a semicircle of radius 5 feet, and it is fixed in a vertical plane with AB horizontal and uppermost. A small heavy bead C, of mass 20 ounces, is threaded on the wire and attached to A by means of a fine light string, of length 6 feet. Find the tension of the string and the pressure between the bead and the wire.

16. ACB is a smooth thin wire in the form of a semicircle, and it is fixed in a vertical plane with AB horizontal and the curve uppermost. A small smooth bead C, of mass 3 ounces, is threaded on the wire, and is attached to A by means of a fine light string equal in length to the radius. Another fine string is attached to the bead and passes through a smooth hook fixed at B, supporting at its other end a mass of 12 ounces. Find the tension of the string AC.

17. A small ring P, of no appreciable weight, is capable of sliding freely on a smooth straight piece of stiff wire AB, fixed in a position inclined 40° to the horizontal with A uppermost. To a fixed point O, situated 1 foot vertically below A, is attached a fine light string which passes through the ring and supports at its free extremity a mass of 5 pounds. Find the length of OP in the position of equilibrium, and the pressure between the ring and the wire.

18. ABC is a smooth fixed wire, the portions AB, BC being straight, and B situated at a higher level than A and C. A small smooth ring of given weight, capable of sliding freely on AB, is connected by a fine light string of given length with a small smooth ring of given weight, capable of sliding freely on BC. Show how to obtain the position of equilibrium, the tension of the string, and the pressures between the wire and the rings.

19. Show that the weight of the greatest mass, which a given force can sustain on a smooth inclined plane of given height, is proportional to the length of the plane.

20. Two smooth inclined planes, of equal height, are placed back to back. Two particles, one on each plane, are connected by a fine light string which passes over the common vertex of the planes. Prove that, if the system is in equilibrium, the weights of the particles are proportional to the lengths of the planes on which they respectively rest.

CHAPTER VII.

EQUILIBRIUM OF A PARTICLE IN CONTACT WITH A ROUGH SURFACE OR CURVE.

62. It is assumed that the student is familiar with the laws of friction, as set forth, for instance, in Loney's *Elements of Statics.*

The reaction of a rough curve or surface is not, in general, normal. It may make any angle, not greater than the angle of friction, with the normal on either side.

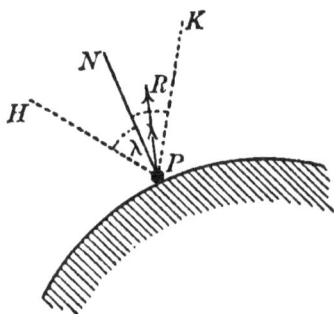
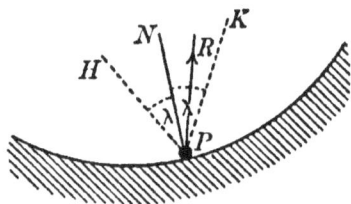

FIG. 68.   FIG. 69.

Suppose a particle rests, in equilibrium, in contact with a rough surface at P. Draw PN, the normal at P, away from the surface, and make angle NPH = angle NPK = the angle of friction = λ. Then the total resistance of the surface, though unrestricted in magnitude must be in a direction intermediate between PH and PK. If it has either of these two limiting directions, the particle is just on the point of slipping.

63. Equilibrium of a Heavy Particle on a Rough Inclined Plane.

A particle of given weight is placed on a rough plane, inclined at a given angle to the horizon, and is sustained by some force applied to it in some direction which is in a vertical plane with the line of greatest slope. The angle of friction between the particle and the plane being given, it is required to consider the conditions of equilibrium.

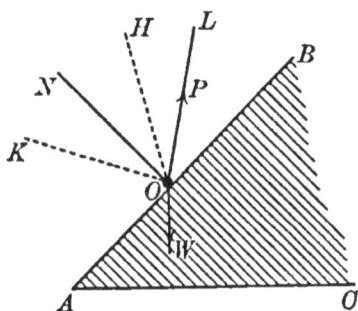

FIG. 70.

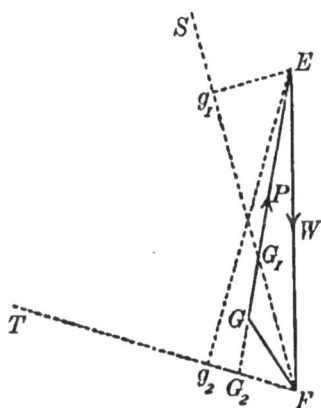

FIG. 70 a.

Let O be the position of the particle, AOB the line of greatest slope making the given angle with the horizontal AC. Suppose the particle is sustained by a force, whose measure is P, applied in direction OL. Let W be the measure of the weight of the particle and let it be represented by EF drawn vertically downwards W units of length.

Let ON be the straight line drawn from O away from the plane and in a direction perpendicular to it. Make angle $NOH =$ angle $NOK =$ angle of friction $= \lambda$; so that OH, OK are within the angles BON, AON respectively. Then the resistance of the plane must be in some direction intermediate between OH and OK, and may be of any magnitude.

Draw FS, FT, straight lines of unlimited length, from F in the directions of OH, OK respectively. The triangle of forces for the particle at O will be a triangle FGE, in which G is some point within the angle SFT, and GE is parallel to OL and of length P units. If G lies in the line FS, the particle is only just prevented from slipping *down* the plane, if in the line FT, it is on the point of slipping *up* the plane.

If the direction of the sustaining force be some given straight line OL, we draw EG_1G_2 parallel to LO to meet FS and FT in G_1 and G_2 respectively. Measure G_1E, G_2E; let them be respectively P_1, P_2 units of length. Then, for equilibrium, P must lie between P_1 and P_2.

Draw Eg_1, Eg_2 perpendiculars to FS, FT respectively. Then g_1E represents in magnitude and direction the smallest force that can prevent the particle from slipping *down* the plane, and g_2E represents the smallest force necessary to drag the particle *up* the plane. Now OH, OK make angles λ with the *normal* to the plane, therefore g_1E, g_2E, which are perpendicular respectively to OH, OK, must make angles λ with the plane itself. Thus, to prevent the particle from slipping *down* the plane, the end is achieved with the least exertion by applying the force in a direction making the angle of

friction with the plane, measured *downwards* from
OB; and to drag the particle *up* the plane, the end is
achieved with the least exertion by applying the force
in a direction making the angle of friction with the
plane, measured *upwards* from *OB*.

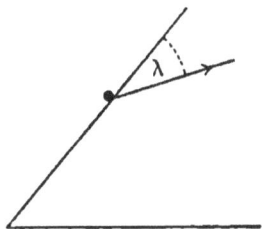

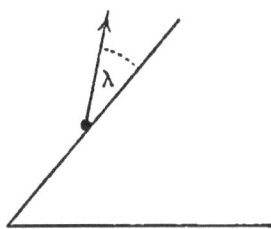

FIG. 71.   FIG. 72.

The left-hand figure shows the easiest way of pre-
venting the particle from slipping *down*; the right-hand
figure shows the easiest way of dragging the particle *up*.

In the space diagram, draw *OM* vertically upwards,
and produce *HO* and *KO* to *H'* and *K'* respectively.

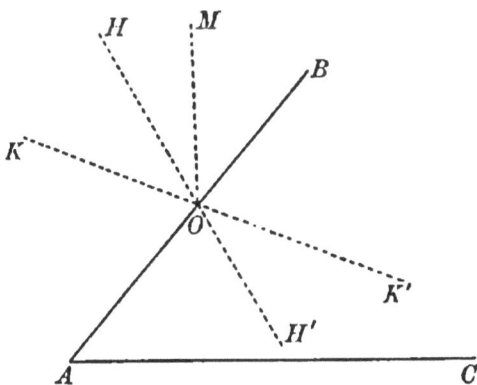

FIG. 73.

Then an examination of the force diagram will show the
student that, for equilibrium, the force *P* must be applied
between the directions *OH'* and *OM*, and that no force,

however great, applied between the directions OH' and OK' can drag the body *up* the plane.

64. Throughout the above piece of work we have taken the inclination of the plane as greater than the angle of friction. The consideration of the case in which the inclination of the plane is less than the angle of friction is left as an exercise for the student. He will see that, in this case, E lies within the angle TFS, and equilibrium is possible for all directions of P. It will be seen that

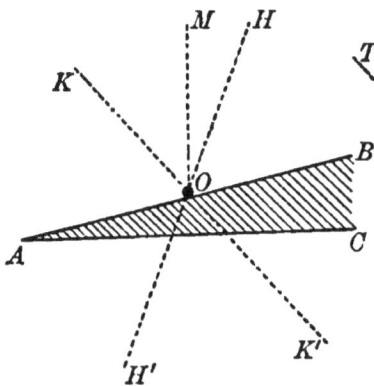

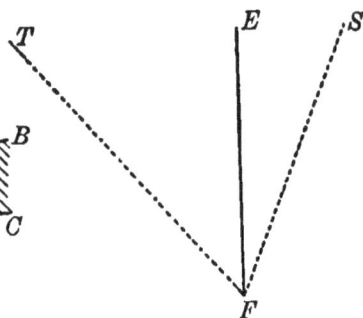

FIG. 74.   FIG. 74 a.

no force applied between the directions OH' and OK' can move the particle. To move it up the plane, the force must be applied between the directions OK' and OM; to move it down, between the directions OH' and OM.

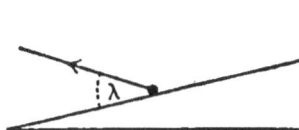

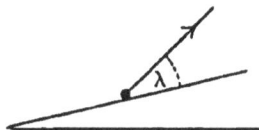

FIG. 75.   FIG. 76.

Also the left-hand figure shows the easiest way of moving the particle *down* the plane; the right-hand figure the easiest way of moving it *up*.

As E lies within the angle TFS, the particle will rest in equilibrium if P vanishes; then the total resistance of the plane is represented by FE, *i.e.* the resistance is equal and opposite to the weight of the particle.

If the inclination of the plane is equal to the angle of friction, FS coincides with FE. In this case, if P vanishes, the particle is just on the point of slipping down the plane.

Conversely, if the particle is just on the point of slipping down the plane under the action of its weight and the resistance of the plane *only*, the angle of inclination of the plane to the horizon is equal to the angle of friction. This gives a method for determining the angle of friction experimentally.

65. *It is required to find at what point, or points, of a given rough curve, fixed in a vertical plane, a particle of given weight may rest* in limiting equilibrium, *being supported by a given force applied to it in a given direction.*

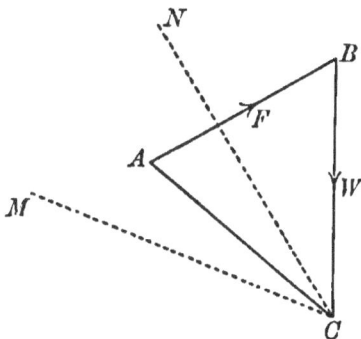

FIG. 77.

Draw AB, BC to represent in magnitude and direction the given sustaining force and the weight of the particle respectively. Then CA must represent the total resistance of the curve, and, as the equilibrium is limiting, CA must make with the normal to the curve an angle equal to the angle of friction. Hence draw CN and CM, making angles equal to the angle of friction on either side of CA. Then, to find the position of the particle in limiting equilibrium, we must find what point, or points, of the curve have their normals in directions CM or CN.

66. Ex. 1. *A particle, of mass* 16 *ounces, rests on a rough plane inclined at an angle of* 30° *to the horizon.*

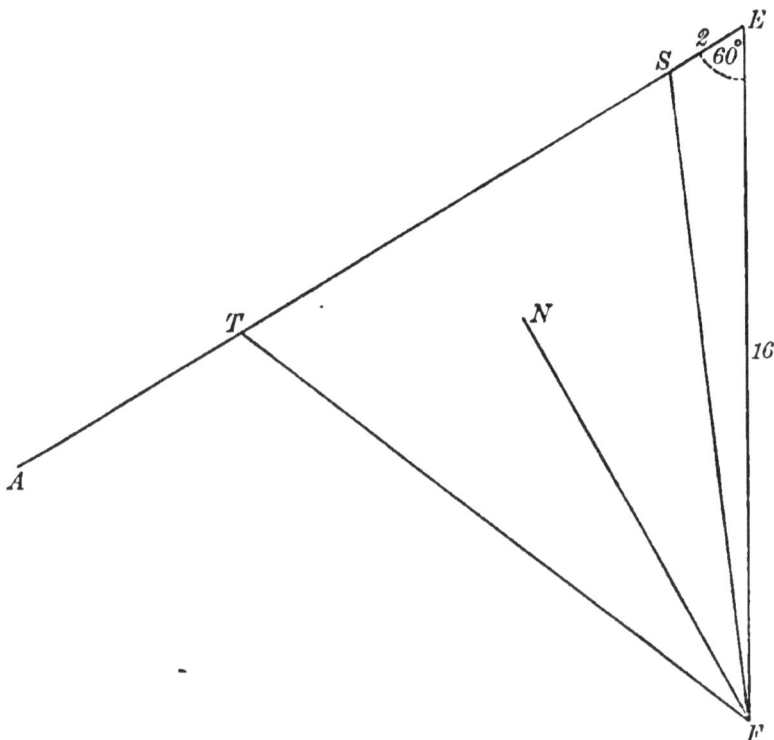

Fig. 78.

If a force equal to the weight of 2 *ounces, acting up and parallel to the plane, is just sufficient to prevent the particle from slipping down, find the force acting in the same direction, which will cause the particle to be on the point of moving up the plane.*

Draw *EF* vertically downwards 16 units of length to represent the weight of the particle. Draw *EA* so that the angle *FEA* is 60°; then *AE* is parallel to the line of greatest slope on the inclined plane. Take *S* in *AE* so that *SE* is of length 2 units, and join *FS*. Draw *FN* perpendicular to *AE*. Then angle *NFS* is the angle of friction. Make angle *NFT* equal to angle *NFS*, and let *FT* meet *EA* in *T*. Then *TE* represents the force required.

We find that *TE* is 14 units of length; therefore the force required is the weight of 14 ounces.

67. *Ex. 2. A particle A, of given weight, is placed upon a given rough inclined plane AD. It is required to determine at what different points of the plane it can be supported by a fine light string ABC, of unlimited length, which passes over a small smooth peg B, situated in a given position, and supports at its other extremity another particle C of given weight. The plane of BAD is vertical, and the line AD a line of greatest slope of the inclined plane.*

The triangle of forces for the particle *A* will be *OHPO* (this way round), in which *OH* is vertically downwards and represents the weight of the particle *A*, *PO* is parallel to *AB*, and represents the weight of the particle *C*, and *HP* represents the total resistance of the plane.

Having drawn *OH*, the point *P* lies on the circumference of a circle having its centre at *O*, and its radius of such length as to represent the weight of the particle *C*.

D.S.   H

Through O draw SOT parallel to AD, to meet in
S and T straight lines drawn through H so as to be
equally inclined to ST, and to contain an angle equal
to twice the angle of friction.   Then, for all possible

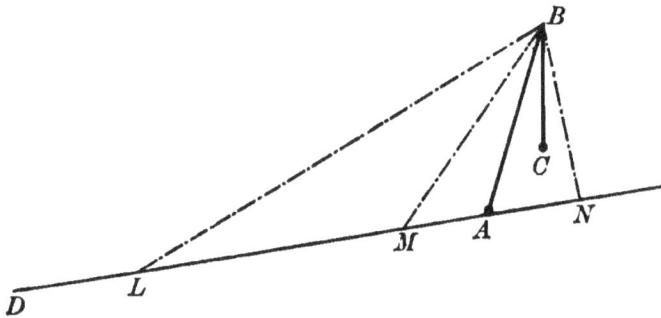

FIG. 79.

positions of equilibrium, the point P must lie within
the angle SHT.   Also, as B is situated *above* the plane,
the point P must lie *below* the line SOT.

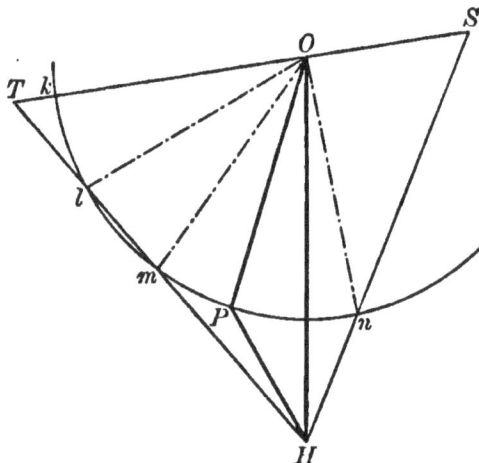

FIG. 79 a.

Hence all possible positions of P are confined to
those portions of the circumference of the circle which
are contained within the triangle SHT.

In the case considered in the accompanying diagram, the only portions of the circumference of the circle contained within the triangle SHT are the arcs kl, mn. Draw BL, BM, BN parallel to Ol, Om, On respectively, to meet AD in L, M, N respectively. Then equilibrium is possible if A lies anywhere between M and N, or anywhere below L.

The student will have no difficulty in interpreting any other case which may occur.

68. Ex. 3. *Two equal heavy particles, on two equally rough inclined planes of the same height placed back to back, are connected by a fine light string which passes over the smooth top edge of the planes; show that, if the particles are on the point of moving, the difference of the inclination of the planes is double the angle of friction.*

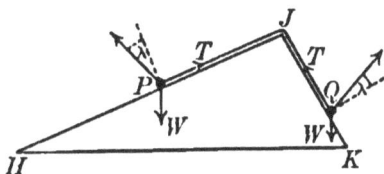
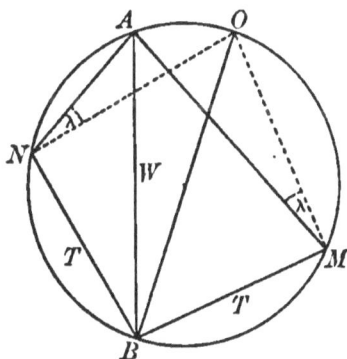

FIG. 80.   FIG. 80 a.

Let P and Q be the particles, each of weight W, resting on the planes HJ, JK respectively, and just on the point of slipping in the direction HJK. Let T be the measure of the tension of the string.

Take AB vertically downwards to represent W, and let BM, BN, parallel to HJ, KJ respectively, be each of length T units. Then MA and NA represent the resistances of the planes at P and Q respectively. Hence NA must be inclined to the normal at Q at an angle λ measured upwards from the outward normal, and MA must make angle λ with the normal at P measured downwards, λ being the angle of friction.

Draw MO, NO perpendiculars to BM, BN respectively. Then the angles OMA, ONA each equal λ. Hence the circle on OB as diameter passes through M, N, A. Since $BM = BN$, the arc $BM = $ arc BN,

$$\therefore \quad \text{arc } MO = \text{arc } NO.$$

$$\therefore \quad \text{arc } MA - \text{arc } NA = \text{twice arc } AO.$$

$$\therefore \quad \text{angle } ABM - \text{angle } ABN = 2\lambda,$$

i.e., the difference of the inclination of the planes to the vertical $= 2\lambda$.

EXAMPLES VII.

1. A particle, of mass 10 ounces, rests on a rough *horizontal* plane, the angle of friction between the particle and the plane being 25°. Find

(i.) the least horizontal force which will move the particle, and determine the total resistance when this force is applied ;

(ii.) the least force which, acting in an upward direction at an angle of 15° with the horizontal, will move the particle, and determine the total resistance ;

(iii.) the magnitude and direction of the total resistance, when a force of 4 ounces' weight is applied in an upward direction making an angle of 20° with the horizontal ;

(iv.) the magnitude and direction of the least force necessary to move the particle.

2. A particle, of mass 10 ounces, rests on a rough plane inclined at an angle of 35° to the *vertical*, the angle of friction between the particle and the plane being 25°. Find

(i.) the least horizontal force which will move the particle *up* the plane;

(ii.) the least horizontal force which will prevent the particle from slipping *down* the plane;

(iii.) the magnitude and direction of the least force necessary to move the particle *up* the plane;

(iv.) the magnitude and direction of the least force necessary to prevent the particle from slipping *down* the plane.

3. A particle, of mass 10 ounces, rests on a rough plane inclined at an angle of 15° to the horizontal, the angle of friction between the particle and the plane being 25°. Find

(i.) the least force which will produce motion when acting *up* the plane;

(ii.) the least force which will produce motion when acting *down* the plane;

(iii.) the magnitude and direction of the least force necessary to move the particle *up* the plane;

(iv.) the magnitude and direction of the least force necessary to move the particle *down* the plane.

4. A particle, of mass 19 ounces, is placed on a rough plane of height 5 feet and length 13 feet, the coefficient of friction being $\frac{1}{2}$; find the magnitude of the horizontal force which will just suffice to push the particle *up* the plane.

What is the magnitude of the horizontal force which will just suffice to drag the particle *down*?

5. A particle, of mass 3 pounds, is just supported on a rough inclined plane, whose height is three-fifths of its length, being acted upon by no forces other than its weight and the resistance of the plane. Find the coefficient of friction between the particle and the plane, and determine the magnitude of the force which, acting parallel to the plane, will be just on the point of moving the particle *up* the plane. Find also the magnitude and direction of the *least* force necessary to move the particle *up* the plane.

6. A is the lowest point of a rough circular hoop fixed in a vertical plane. A small ring, of mass 2 pounds, is threaded on the hoop at P, where the arc AP subtends 50° at the centre of the hoop. Find the magnitude of the smallest horizontal force which will support the ring, the angle of friction between the ring and the hoop being 20°.

7. A particle, of weight W, is sustained in limiting equilibrium on a rough circular hoop, fixed in a vertical plane, by a force $\frac{1}{2}W$ acting at an angle of 60° with the vertical and upwards. The angle of friction between the particle and the hoop being 20°, find the positions of limiting equilibrium.

8. A ring P, of inappreciable weight, is capable of sliding on a rough straight piece of wire ACB, which is fixed in a position inclined at an angle of 65° to the vertical. A fine light string has one extremity attached to the ring P, passes through a small smooth ring D, fixed at a distance of 10 inches vertically below C, and supports at its free end a mass of weight W. If the coefficient of friction between the ring and the wire is $\frac{2}{3}$, show that there is a portion of wire, of length about $7\frac{1}{4}$ inches, at any point of which the ring can rest in equilibrium, and that beyond either end of this portion equilibrium is impossible.

9. A body is placed on a rough inclined plane. Prove that the force which must be applied to it in a fixed direction, in order to just prevent it from slipping *down*, is the same as if the plane were made smooth and its inclination to the horizon were decreased by the angle of friction.

10. A heavy body is supported on a rough plane inclined at an angle 2λ to the *vertical*, λ being the angle of friction. It is just on the point of moving *up* the plane when acted upon by a force parallel to the plane. Show that the applied force must be equal to the weight of the body.

11. A particle is supported on a rough inclined plane by a force equal to the weight of the particle. Show that the force must be applied in some direction within a fixed angle equal to four times the angle of friction.

12. A particle, placed on a rough inclined plane, is on the point of slipping down the plane, being acted upon by no force other than its weight and the resistance of the plane. Show that the least force which, acting parallel to the plane, will move it up the plane, is twice as great as the force which would support it, if the plane were smooth. Prove also, that the total resistance of the plane, in the first case, is equal to the weight of the particle.

13. A particle, of weight W, is supported on a rough inclined plane by a force acting up the plane. It is on the point of moving up the plane when the force has the value P_1, and of moving *down* the plane when the force has the value P_2. Show that a force $\frac{1}{2}(P_1+P_2)$, acting up the plane, would support the same particle on a smooth plane of the same inclination.

14. A heavy particle is placed on a rough *horizontal* plane, the angle of friction between the particle and the plane being λ. When the particle is on the point of moving, under the influence of a horizontal force, the total resistance of the plane is of magnitude R. Show that the force, which, applied at an angle 3λ with the downward vertical, is just sufficient to move the particle, is of magnitude $\frac{1}{2}R$.

15. Show how to determine all possible positions of equilibrium of a heavy bead on a rough circular wire, which is fixed in a vertical plane.

16. A small bead, of no appreciable weight, is capable of moving on a circular wire, fixed in a vertical plane. A fine light string, attached at one extremity to the bead, passes over a small smooth peg, situated at a point on the wire, and supports at its other extremity a heavy body. Show that the bead can rest in equilibrium at any point of a particular arc, which subtends, at the centre of the wire, an angle equal to four times the angle of friction.

17. The triangle ABC, right-angled at C, is the vertical section o a rough inclined plane. When the plane is placed with AC horizontal and BC vertical, a certain force of unknown magnitude X, acting parallel to the plane, can just move a mass of given weight W_1 *up* the plane; when the plane is placed with BC hori-

zontal and AC vertical, the same force X, acting parallel to the plane, can just prevent a mass of given weight W_2 from moving *down* the plane. Prove the following construction for determining the angle of friction, which is the same for both masses, and the value of X :

Take LH and LK at right angles, to represent W_1 and W_2 respectively, and make angle KLM equal to the angle BAC, so that LM meets HK in M. Draw LN perpendicular to HK. Then LM represents X, and MLN is the angle of friction.

18. Prove the following particular cases of the preceding example :

(i.) If $W_1 : W_2 = AC : BC$, the plane is smooth.

(ii.) If $W_1 : W_2 = BC : AC$, the total resistance of the plane, in both positions, and the force X have the same magnitude ; also, half of the angle of friction is equal to the difference between 45° and the angle BAC.

(iii.) If $AC = BC$, then, in turning the plane round, the total resistance is changed in the ratio $W_1 : W_2$.

(iv.) If W_1, W_2 and X are all known, while the inclination of the plane and the coefficient of friction are to be determined, there may be two planes of different inclinations, but of the same coefficient of friction, satisfying the given conditions.

19. The triangle ABC, right-angled at C, is the vertical section of a rough inclined plane. When the plane is placed with AC horizontal and BC vertical, a given force P, acting parallel to the plane, can just move a mass of unknown weight W *up* the plane ; when the plane is placed with BC horizontal and AC vertical, another given force Q, acting parallel to the plane, can just prevent the same mass from slipping *down* the plane. Prove the following construction for determining the weight of the mass and the angle of friction between the mass and the plane :

Draw LH and LK at right angles, to represent P and Q respectively, and make angle KLM equal to the angle BAC, so that LM meets the circle HLK in M. Draw the diameter LN of this circle. Then LM represents W, and the angle NLM is the angle of friction.

20. Prove the following particular cases of the preceding example :

(i.) If $P : Q = BC : AC$, then the plane is smooth, and, in turning it round, the total resistance is changed from Q to P.

(ii.) If $P : Q = AC : BC$, then, in turning the plane round, the total resistance is changed from P to Q ; also, half of the angle of friction is equal to the difference between 45° and the angle BAC.

(iii.) If P, Q and W are all known, while the inclination of the plane and the coefficient of friction are to be determined, there may be two planes of different inclinations, but of the same coefficient of friction, satisfying the given conditions.

CHAPTER VIII.

TWO FORCES WHOSE LINES OF ACTION DO NOT INTERSECT AT AN ACCESSIBLE POINT.

69. *To find the resultant of two given forces, when the point of intersection of their lines of action is inaccessible.*

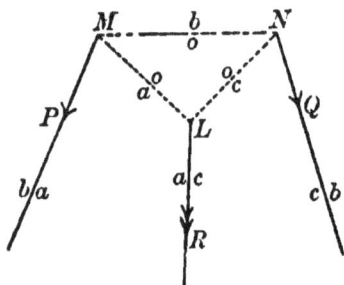

FIG. 81.

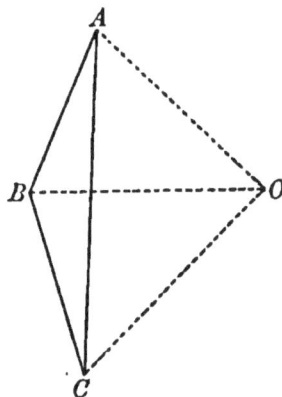

FIG. 81 *a*.

Let two forces, whose measures are P and Q, act along two given lines. Draw AB of length P units in the direction of the force P, and BC of length Q units in the direction of the force Q. Then AC represents the resultant in magnitude and direction. If AC is found to contain R units of length, R is the measure of the resultant. We have to find its line of action.

Take any two points, M and N, in the lines of action of P and Q respectively. Draw BO parallel to MN, and, taking any point O in BO, join A, O and C, O. Draw ML and NL parallel to AO and OC respectively, to meet in L. Then we shall show that L is a point in the line of action of the resultant.

The force P can be replaced by two forces represented by AO, OB acting in the lines ML, NM respectively; and the force Q can be replaced by two forces represented by BO, OC acting in the lines MN, NL respectively. Let P and Q be replaced by these pairs of components. Then the two forces represented by OB and BO acting in the line MN balance one another, having no effect on the body as a whole, and may therefore be removed.

Hence the two given forces have the same resultant as forces represented by AO, OC acting in the lines ML, NL respectively.

Hence a straight line drawn through L parallel to AC is the line of action of the resultant, and the measure of the resultant is R.

The correspondence between the two figures will be made clearer if we use the notation suggested in Art. 4. The two forces P and Q are represented by AB and BC respectively; we therefore denote their lines of action by ab and bc respectively, placing one of the small letters on each side of the line indicated. The straight line MN, connecting a point in ab with a point in bc, is denoted by ob, and the line OB of the force diagram is parallel to it. Through the intersection of ab and bo is drawn ao parallel to AO, and through the intersection of bc and bo is drawn oc parallel to OC. The intersection of *oc* and *ao* gives us a point in *ac*, which is the line of action of the resultant represented by *AC*.

The above method is, of course, applicable to the case in which the lines *bc* and *ab* intersect at an accessible point. We know that the line of action of the resultant passes through that point. It can be proved geometrically that the line *ac* passes through the point of intersection of *ab* and *bc*.

70. Parallel Forces.

The construction for the resultant of two parallel forces is a particular case of the above.

Case I. Let the two forces *P* and *Q* be parallel and in the same direction. Making the construction of the preceding article, we see that the resultant is equal to the sum of the forces, and is in the same direction

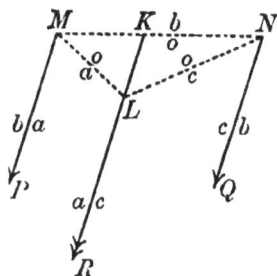
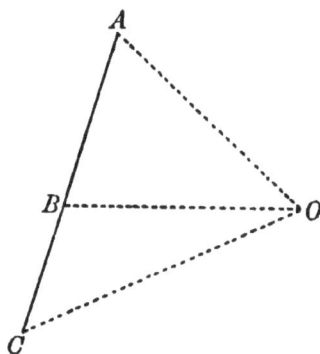

FIG. 82.   FIG. 82 *a*.

as each of its components. Also, if the line of action of the resultant meets *MN* in *K*, we have

$$\frac{MK}{KN} = \frac{MK}{KL} \cdot \frac{KL}{KN} = \frac{OB}{BA} \cdot \frac{BC}{OB} = \frac{BC}{BA} = \frac{Q}{P}.$$

∴ the point *K* divides *MN* internally in the inverse ratio of *P* to *Q*. If we make use of this result, we can dispense with the force diagram altogether, or we may use it to obtain the following more simple construction:

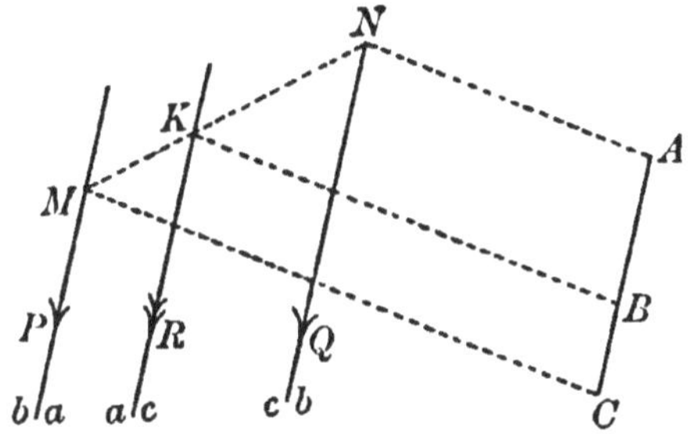

FIG. 83.

Join A to any point N in the line bc. Draw CM parallel to AN to meet the line ab in M. Draw BK parallel to AN to meet MN in K. Then K is a point in the line of action of the resultant.

Case II. Let the two forces P and Q be parallel and in opposite directions.

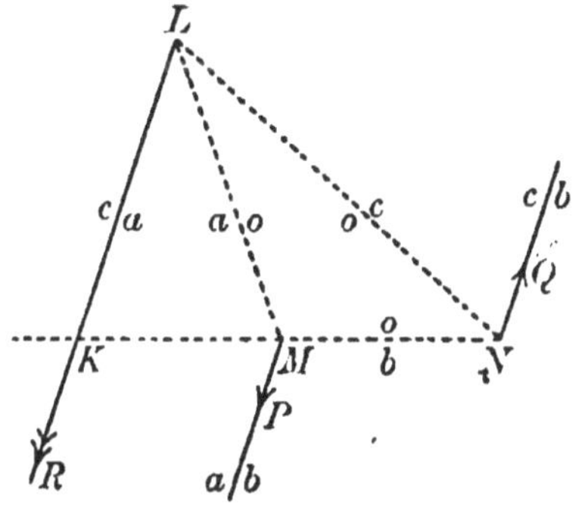

FIG. 84. 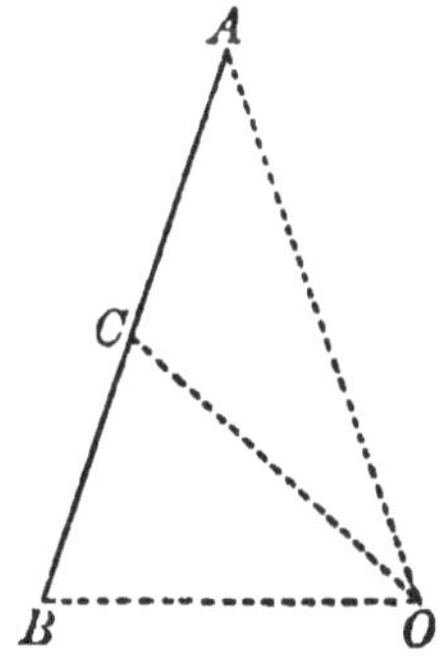   FIG. 84 a.

Making the same construction as before, we see that the resultant is equal to the difference of the two given forces, and acts in the direction of the greater. (In the figure we have taken $P > Q$.)

· Also, if the line of action of the resultant meets NM produced in K, we have

$$\frac{MK}{KN}=\frac{MK}{KL}\cdot\frac{KL}{KN}=\frac{OB}{BA}\cdot\frac{BC}{OB}=\frac{BC}{BA}=\frac{Q}{P}.$$

∴ the point K divides MN *externally* in the inverse ratio of P to Q.

We see that the above construction fails if P is equal and opposite to Q. In that case C coincides with A, and the straight lines oa, oc become parallel, so that there is no point L at a finite distance. Two equal parallel forces acting in opposite directions therefore have no resultant; they are said to form a *couple*.

The result obtained above gives the same simplified construction for finding the resultant of two parallel forces in opposite directions as for finding the resultant of two parallel forces in the same direction. The figure is shown below:

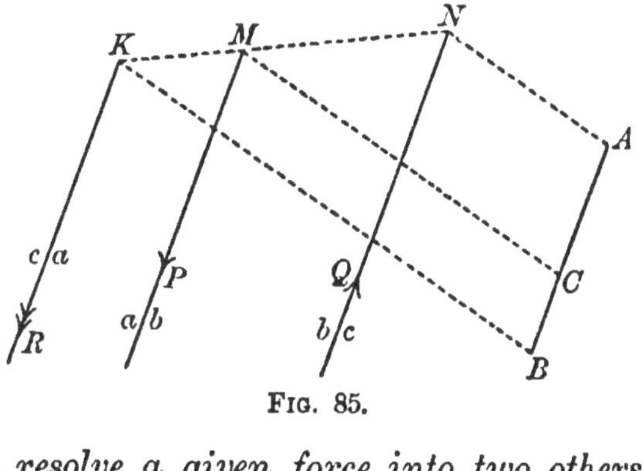

FIG. 85.

71. *To resolve a given force into two others passing through two given points.*

Let the given force be represented by AB, and let its line of action be marked ab. Let H and K be the given points.

From H and K draw straight lines oa, ob respectively, to meet at any point chosen on ab; draw AO, BO parallel to ao, bo respectively, to meet in O. Join H, K, and let the straight line so drawn be called ox. Draw OX parallel to ox, and in OX take *any* point X.

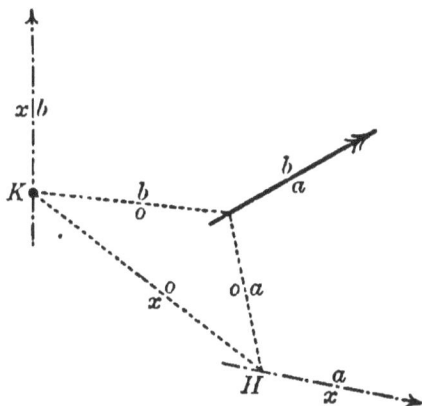

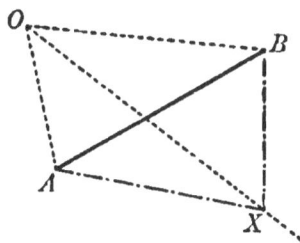

FIG. 86.   FIG. 86 a.

Then AX and XB represent components of the given force acting through the points H and K respectively. As X may be taken anywhere in the line through O drawn parallel to HK, the problem is indeterminate.

If the two components are required to be parallel, we take X in AB; if equal, we take X in the straight line which bisects AB at right angles. In the former case the lines of action of the components are each parallel to ab.

72. *To resolve a given force into two others acting along two given straight lines each parallel to the line of action of the given force.*

This is a particular case of the preceding article, and can be solved in the manner there indicated. We choose the points H, K anywhere in the lines of action

of the required components respectively, and resolve
the given force into two parallel forces acting through
the points H, K.

Or we may proceed thus:

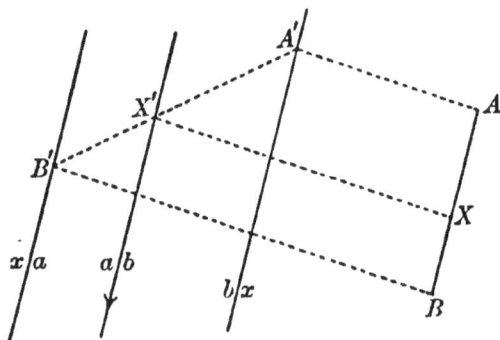

FIG. 87.

Let the given force be represented by AB, and let
its line of action be marked ab. Of the lines of action
of the required components mark one ax, the other xb.
Draw two parallel lines through A, B to meet bx, ax
in A', B' respectively. Let $A'B'$ meet ab in X', and
draw $X'X$ parallel to $A'A$ to meet AB in X. Then,
clearly, $AX : XB = A'X' : X'B'$, and therefore AX and
XB represent the components required along ax and
xb respectively.

We have drawn the figures for the case in which
the given force lies *between* the lines of action of its
components. The method is, however, quite general.

We might dispense with the force diagram altogether,
and proceed as follows:

Draw any straight line $A'X'B'$ meeting the lines
bx, ab, xa in the points A', X', B' respectively.

Measure $A'X'$ and $X'B'$, and divide the given force into

two parts in the ratio $A'X':X'B'$. The first part is the component along ax, the other that along xb.

73. Ex. 1. *Forces of 24 and 10 pounds' weight act along the straight lines MH and NK respectively; the angles HMN and MNK are 90° and 100° respectively, and the points H, K are on the same side of MN, which is 10 inches long. Find the magnitude and direction of the resultant of the forces, and determine the point where its line of action cuts MN. Find also the resultant when the force of 10 pounds' weight is reversed in direction.*

Draw AB, BC in the directions of MH, NK respectively, and of lengths 24 and 10 units respectively. In the straight line drawn through B parallel to MN take any point O. Draw ML, NL parallel to AO, OC respectively, to meet in L. Let the straight line drawn through L parallel to AC meet MN in X. Then XL is the line of action of the resultant, which is represented by AC.

On measurement, we find that AC is 33·9 units of length, and that the angle MXL is 93°. Hence the resultant is 33·9 pounds weight, and makes an angle of 93° with NM. Also MX is found to be 2·9 units of length; therefore the resultant acts through a point X in MN distant 2·9 inches from M.

Produce CB to C', making $BC'=CB$. Then, if the force of 10 pounds' weight be reversed, it will be represented by BC'. Draw NL' parallel to OC' to meet LM produced in L', and $L'X'$ parallel to AC' to meet NM produced in X'. Then $L'X'$ is the line of action of the resultant, which is represented by AC'.

On measurement, we find that the resultant is now

D.S.   I

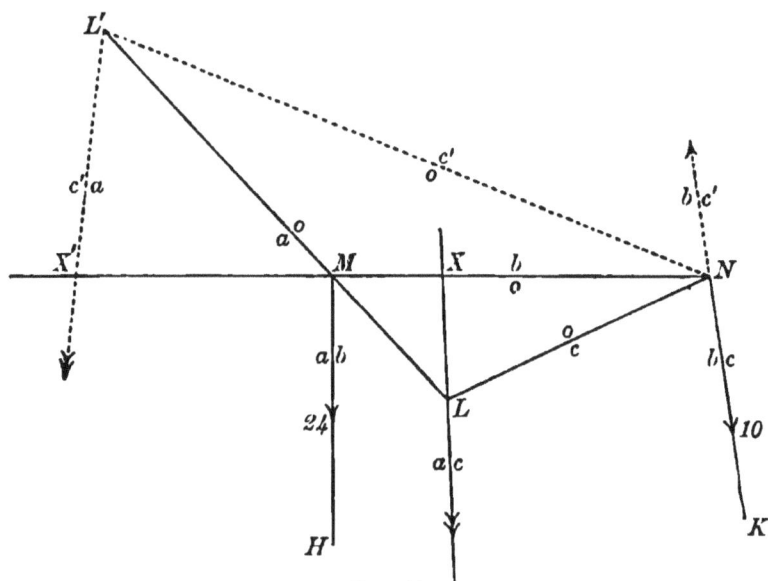

FIG. 88.

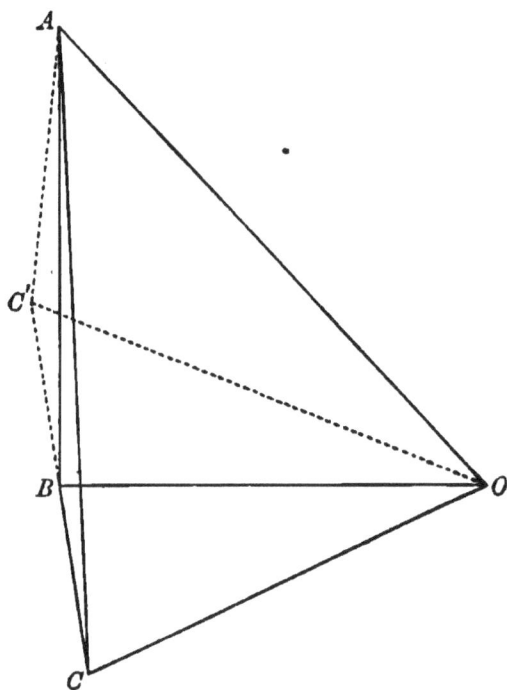

FIG. 88 a.

14·3 pounds' weight in a direction 96° with MN, and that it acts through a point X' in NM produced, distant 6·9 inches from M.

74. Ex. 2. *If from any point, in the line of action of the resultant of two forces, perpendiculars be drawn upon the lines of action of those forces, the lengths of the perpendiculars are inversely proportional to the magnitudes of the forces.*

In Art. 69 the position of the point O is arbitrary. Let O be taken at the other extremity of the diameter through B to the circle described passing through A, B, C. Then OA, OC are perpendicular to BA, BC respectively; therefore LM, LN are the perpendiculars from L upon the lines of action of the forces P and Q respectively.

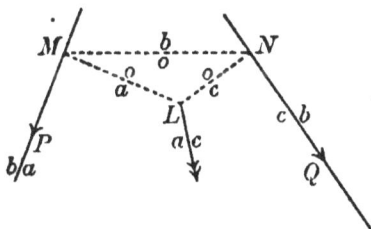

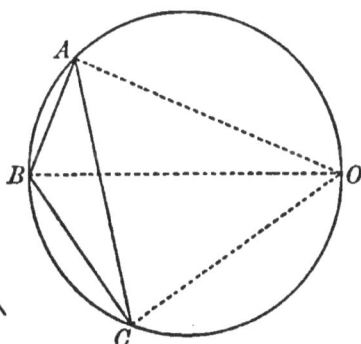

FIG. 89.   FIG. 89 a.

Now   angle $NML =$ angle $AOB =$ angle ACB,
and   angle $MNL =$ angle $COB =$ angle CAB.
∴   the triangle MNL is similar to the triangle CAB.
∴ $LM : LN = BC : BA = Q : P$.

Now L is a point in the line of action of the resultant; also MN may be taken in any position, provided only that it is parallel to BO; therefore, as MN moves, remaining parallel to BO, the point L traces out the line of action of the resultant.

Therefore, from any point on the line of action of the resultant, the perpendiculars let fall upon the lines of action of P and Q are proportional to Q and P respectively.

EXAMPLES VIII.

1. Two parallel forces $5P$ and $7P$ act at points A and B respectively. Find the magnitude, direction, and position of their resultant (i.) when the forces are *like*, and (ii.) when *unlike*.

2. Two parallel forces, of 20 and 25 pounds' weight and of opposite senses, act on a rigid body, the perpendicular distance between their lines of action being 4 inches ; find their resultant.

3. $ABCD$ is a square, and E is taken in AD so that $AE=\frac{1}{3}AD$. Find the magnitude and line of action of the resultant of two forces, one of which is 20 pounds' weight acting at E in direction EB, and the other is 15 pounds' weight acting at D in direction DC.

4. Assuming the magnitude, direction, and position of the resultant of two *like* parallel forces, deduce the magnitude, direction, and position of the resultant of two *unlike* parallel forces.

5. Show how to find the magnitude of a force acting along a given line, in order that the resultant of this force, and a second force given in magnitude, direction, and position, may pass through a given point.

6. Show how to resolve a given force into two others, one of which is along a given line of action, and the other of which passes through a given point.

7. If, in the figure of Art. 69, the lines ac, ob intersect in K and AC, OB in X, then $MK : KN = CX : XA$.

8. Two forces, whose magnitudes are in a given ratio, act at points A and B respectively of a rigid body. Prove that, what-

ever be the directions of the forces, provided only that they are *either* parallel to one another *or* equally inclined to AB, the line of action of their resultant always passes through a fixed point in AB.

9. Show how to resolve a given force into two others which act each through a given point and are in a given ratio.

10. Two forces P and Q act at two given points M and N respectively, and the line of action of their resultant meets MN in K. If $P : Q = NK : KM$, prove that *either* P and Q are parallel, *or* they are equally inclined to MN.

11. Two forces are represented in magnitude, direction, and position by the lines AB, CD; H and K are the middle points of AC, BD respectively; straight lines through A and C parallel to BH, DH respectively meet in h, and straight lines through B and D parallel to AK, CK respectively meet in k. Prove that hk is the line of action of the resultant of the two forces.

In particular, show how to make use of the above in determining the line of action of the resultant of two given parallel forces.

.

CHAPTER IX.

EQUILIBRIUM OF THREE FORCES ACTING IN ONE PLANE UPON A RIGID BODY.

75. *Three forces act upon a rigid body in one plane; it is required to consider the conditions of equilibrium.*

Let P, Q, R be the measures of three forces acting in one plane upon a rigid body along the lines bc, ca, ab respectively, and keeping it in equilibrium.

First method.

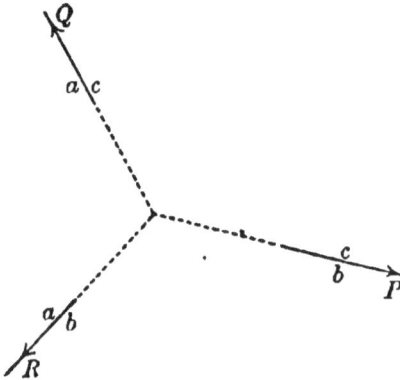

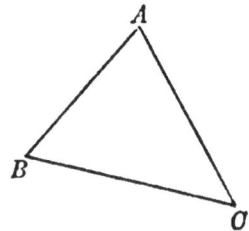

FIG. 90.   FIG. 90 a.

Let BC, CA be drawn in the directions of P and Q and of length P units and Q units respectively. Then the resultant of P and Q is represented by BA, and acts through the point of intersection of bc and ca. Let P and Q be replaced by this resultant. Then we have two forces keeping a rigid body in equilibrium. These two forces must be equal and act in opposite directions along the same straight line. Therefore AB is the direction of R and of length R units, and ab, the line of action of R, must pass through the point of intersection of bc and ca. This point will be called the point abc.

Hence, if three forces acting in one plane upon a rigid body keep it in equilibrium, their lines of action must be concurrent, and the force diagram is a triangle whose sides represent the forces taken one way round.

76. *Second method.*

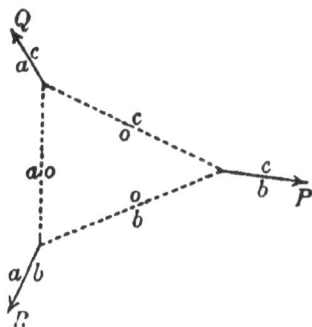 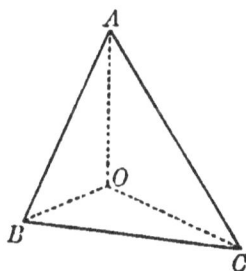

FIG. 91.   FIG. 91 *a*.

As before, let BC, CA be drawn to represent P and Q in direction and magnitude, and join AB. Take any point O in the force diagram, and connect it with the points A, B, C.

From a point in the line bc draw straight lines ob, oc parallel to OB, OC respectively. From the point of intersection of oc and ac draw oa parallel to OA. We shall show that the lines oa, ob, ab are concurrent.

Let a, β, γ be the measures of OA, OB, OC respectively. Then the force P may be replaced by two forces β and γ acting along the lines bo, oc in the directions BO, OC respectively. Also the force Q may be replaced by two forces γ and a acting along the lines co, oa in the directions CO, OA respectively. Let the two forces P and Q be replaced by these pairs of components. Then the two forces γ, acting in opposite directions along the line oc, balance one another, and may therefore be removed. We are now left with three forces β, a, R in equilibrium. The first two of these are represented by BO, OA, and act along the lines bo, oa respectively. Hence, by the *first method* above, AB must represent R in magnitude and direction, and the three lines ob, oa, ab must be concurrent.

The student should carefully notice the correspondence in the two figures. The straight line in the force diagram from O to the point of intersection of the lines which represent P and Q, is parallel to the straight line in the space diagram which connects points in the lines of action of P and Q, and similarly for other pairs of forces. The two dotted lines which intersect on the line of action of one of the forces P, are parallel to the lines from O to the extremities of the line which represents P in the force diagram.

77. The first of the above two methods is the fundamental method, and we have quoted it in dealing with the second, but it ceases to be of practical use when the point abc is inaccessible. The second is a more general method, and is always applicable, as the point O may be taken anywhere in the diagram, so that no two lines need intersect at very acute angles.

Instead of choosing the point O before drawing the dotted lines in both figures, we may choose any three points, one on the line of action of each of the three forces, and thus draw the triangle whose sides are oa, ob, oc. Then the straight lines through A, B, C parallel to oa, ob, oc respectively must be concurrent. For, take O to be the point of intersection of the straight lines drawn through B and C parallel to ob, oc respectively. Then, by the preceding bit of work, the straight line drawn through the intersection of oc and ac, parallel to AO, must pass through the intersection of ob and ab. That is, ao is parallel to AO.

The triangle whose sides are oa, ob, oc can always be chosen so that the lines cut at convenient angles, and thus we have a method applicable to all cases.

78. *Third method.*

The following method is a particular case of the preceding method, the point O being taken in BC, but it is of sufficient practical importance to be treated separately:

As before, let BC be drawn to represent the force P, and let the lines of action of P, Q, R be marked bc, ca, ab respectively.

Resolve the force P into two parallel forces β and γ represented by BO, OC respectively, acting along two lines bo, oc respectively; so that O is a point in BC, and bo, oc are both parallel to bc.

Let the lines oc, ca intersect at K, and the lines bo, ab at L, and let KL be marked ao.

Replacing the force P by its components β and γ, we see that γ and Q acting at K must balance β and R acting at L. Therefore the resultant of the first pair and the resultant of the second pair must be equal and act in opposite directions along the same straight line ao. Hence, if CA represents Q, then OA, which represents the resultant of γ and Q, must be parallel to oa. Also AO must represent the resultant of R and β, and therefore, as BO represents β, AB must represent R.

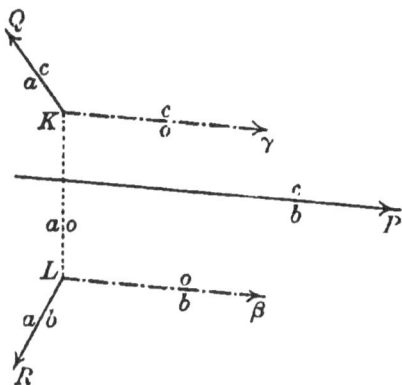

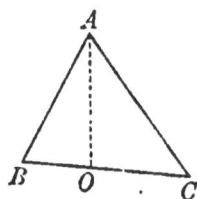

FIG. 92.   FIG. 92 a.

Thus, for equilibrium, it is necessary and sufficient that the straight lines through B, O, C parallel to ba, oa, ca respectively should meet at a point A, and that CA and AB should represent Q and R respectively.

The student should notice that O divides BC in the same ratio that the line bc divides KL. The method becomes practically useful when the points K and L are given, and when the position of the line bc is known relatively to K and L. The point O can then be quickly determined, and the force diagram completed.

79. Parallel Forces.

The equilibrium of three parallel forces is a particular case of the above, the second method alone being applicable.

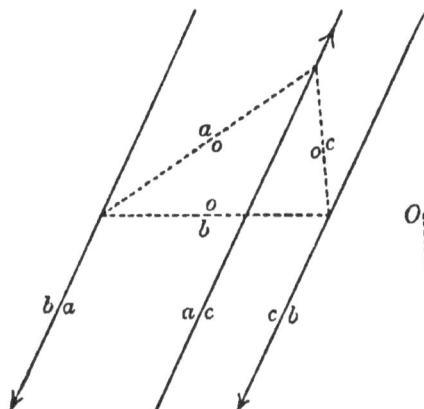

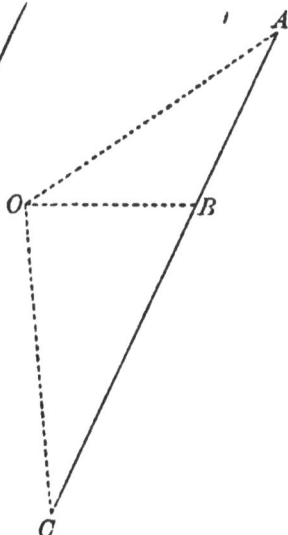

FIG. 93.   FIG. 93 a.

Here the points A, B, C are collinear.

Otherwise.—We may proceed as follows: Let two parallel forces acting along the lines ab, bc be in equilibrium with a third force acting along the line ca. Take AB, BC to represent the two forces acting along ab, bc respectively. Draw any straight line AA' to meet bc in A', and let the straight line through C drawn parallel to AA' meet ab in C'. Draw BB' parallel to AA' to meet $A'C'$ in B'. Then, by Art. 70, B' is a point in the line of action of the resultant of the two forces which act along ab, bc, and this resultant is represented by AC. Hence the system is equivalent to two forces

in equilibrium—the one represented by AC and acting
through B', the other acting along ca. These must be
equal and in opposite directions along the same straight

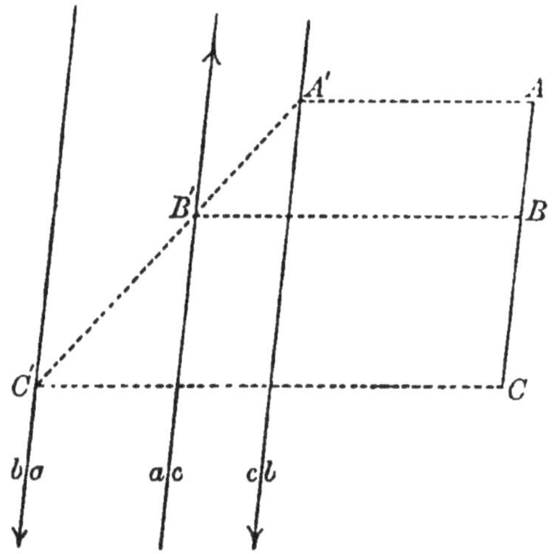

FIG. 94.

line. Hence B' must be a point in ca, and the force
which acts along ca is represented by CA.

80. *A known force, whose measure is P, acts in a
given direction along a given straight line bc; a force
of unknown measure X acts along a given straight line
ca; and a third force of unknown measure Y acts in
an unknown direction through a given point H. It is
required to find the values of X and Y and the direction
of Y, in order that the three may be in equilibrium.*

First method.

Join H to the point of intersection of bc and ca, and
let the straight line thus drawn be called ab. Then
ab is the line of action of Y.

Draw BC in the direction of P and of length P units,
and through B and C let straight lines be drawn parallel

to ba and ca respectively, to meet in A. Then $BCAB$
(this way round) is the force diagram for the system.

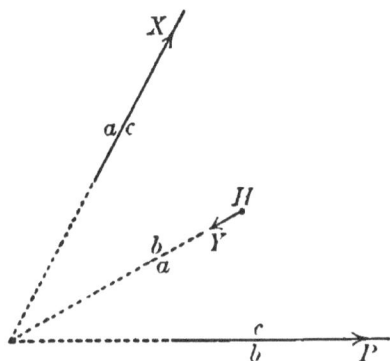

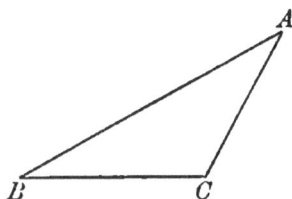

FIG. 95.   FIG. 95 a.

Measuring CA and AB, we have X and Y respectively,
and the direction of Y has already been determined.

This method fails when the point of intersection of
bc and ca is not accessible.

Second method.

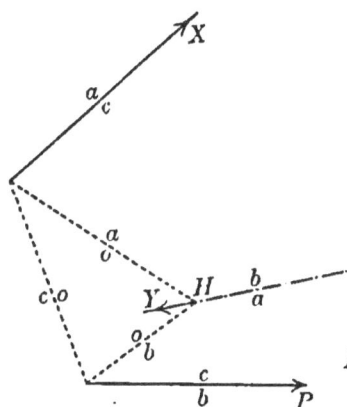

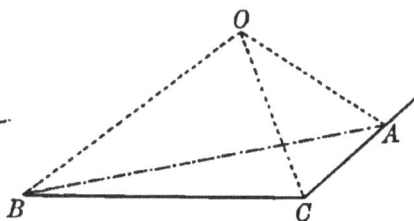

FIG. 96.   FIG. 96 a.

Draw BC in the direction of P and of length P units,
and from C draw CA parallel to the line ca. This

does not determine the position of A, as we do not know X. Draw any straight line oc intersecting the lines bc, ca. Let the straight line connecting H with the intersection of bc and co be drawn and called ob, and let the straight line connecting H with the intersection of ca and co be drawn and called oa. Through B and C draw straight lines parallel to bo and co respectively, meeting in O, and through O draw OA parallel to oa to meet CA in A. This determines the point A. Then, joining AB, and measuring CA and AB, we have X and Y respectively. Also, the line of action of Y is the straight line ab drawn through H parallel to AB.

Third Method.

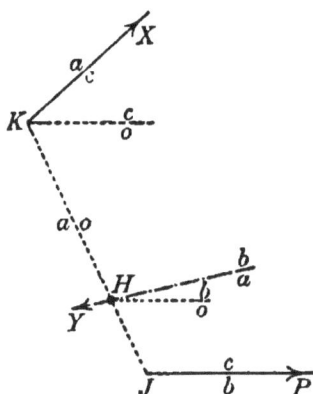
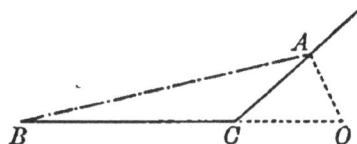

FIG. 97.   FIG. 97 a.

Draw BC in the direction of P, and of length P units, and from C draw CA parallel to the line ca, the point A being at present unknown.

Let a straight line ao, passing through H, meet the lines bc, ca in J and K respectively; and through H and K draw straight lines ob, oc respectively, each parallel to bc. Find a point O in BC such that BO,

OC represent components of P acting in the lines ob, oc respectively. This point O will divide BC similarly to the way in which J divides KH, and the method is practically useful when the relative positions of the points J, H, K are known.

Through O draw OA parallel to oa to meet CA in A. Then, joining AB, and measuring CA and AB, we have X and Y respectively. Also, the line of action of Y is the straight line ab drawn through H parallel to AB.

81. Equilibrium of Four Forces, two of which are fully known.

The methods of this chapter may be applied to the consideration of the equilibrium of four forces, two of which are fully known. We may replace the two known forces by their resultant, thus reducing the number of forces to three, one of which is fully known.

82. Centre of Gravity. In the examples which here follow, it is assumed that the resultant of the weights of the constituent elements of a body (or of any material system in which the parts retain the same position with regard to one another) acts along a vertical line, which always passes through a special point of the body called its centre of gravity, this point retaining the same position with regard to the different parts of the body, in whatever position the body is placed; also, that the centre of gravity of a body of uniform density and symmetrical shape, is in the position of the centre of symmetry.

83. Smooth Hinge. A body is said to be *capable of turning freely* about a fixed point O, or to be *smoothly*

hinged to a fixed point O, when the point O of the body
is compelled to remain in the position of the point O in
space, the constraint being a force acting through O of
the nature of a direct push or pull. Such a force is
self-adjusting, and accommodates itself to prevent the
point O of the body from getting away from the point
O of space, if possible. It is of any magnitude, and acts
in any direction necessary to preserve equilibrium, but
must act through the point O.

84. Ex. 1. *A thin uniform rod AB, of length 10 feet
and weighing 12 pounds, is capable of turning freely
about a smooth hinge at A. It is supported by a fine
light string, of length 10 feet, connecting B with a point
C, situated 16 feet from A in a horizontal line. Find
the tension of the string and the action at the hinge.*

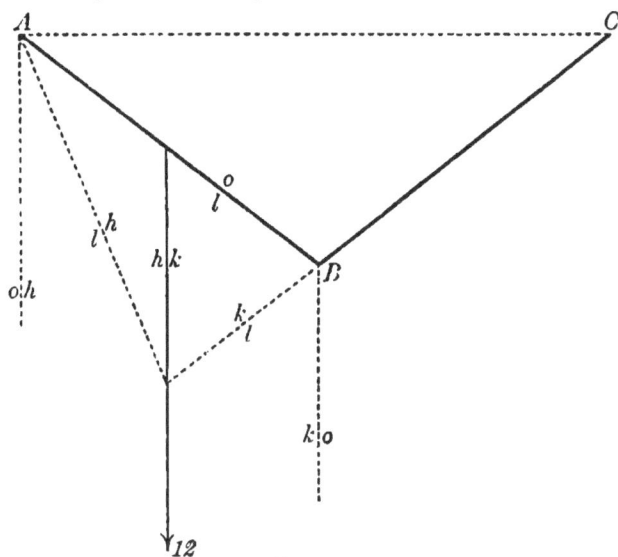

Fɪɢ.

Having constructed the space diagram to scale according to the data, draw the straight line *hk* vertically

downwards through the middle point of *AB*, and a
straight line *HK*, 12 units of length, vertically downwards to represent the weight of the rod. Mark the
line *BC* with the letters *kl*, and
draw the straight line *lh* from
A to the point of intersection
of *hk* and *kl*.

Through *H* and *K* draw
straight lines parallel to *hl* and
kl respectively, meeting in *L*.
Then *KL* represents the tension
of the string, and the action of
the hinge upon the rod is represented by *LH*, and acts in
the line *lh*.

We find that *KL* = 5 units
and *LH* = 9·85 units. Hence
the tension of the string is
5 pounds' weight, and the
reaction at the hinge is
9·85 pounds' weight, and is
found to make an angle of 66°
with *AC*.

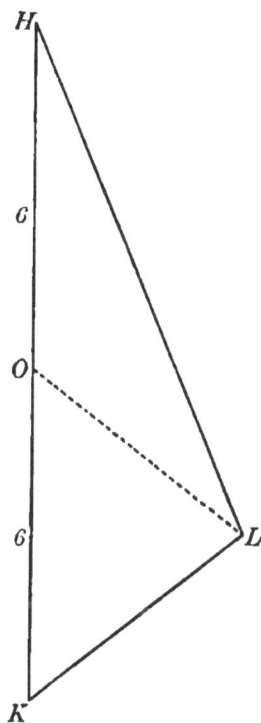

FIG. 98 a.

Otherwise.—Instead of drawing the line *lh*, draw
through *A* and *B* respectively straight lines *oh* and *ok*,
each parallel to *hk*. Bisect *HK* in *O*. Then the weight
of the rod is equivalent to forces represented by *HO*,
OK acting along the lines *ho*, *ok* respectively.

Let *AB* be marked *ol*, and draw *OL* parallel to *ol* to
meet in *L* the straight line drawn through *K* parallel
to *kl*. Then, measuring *KL*, we have the tension of the
string, and the straight line *LH* gives the magnitude

and direction of the action at the hinge.   This construction, of course, gives the same result as before.

85. Ex. 2. *In the above example, let C be situated
4 feet, instead of 16 feet, from A in a horizontal line,
the problem in other respects remaining unaltered.*

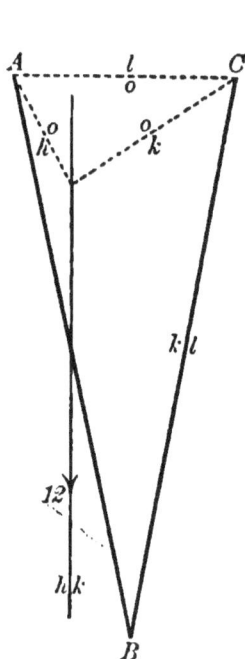
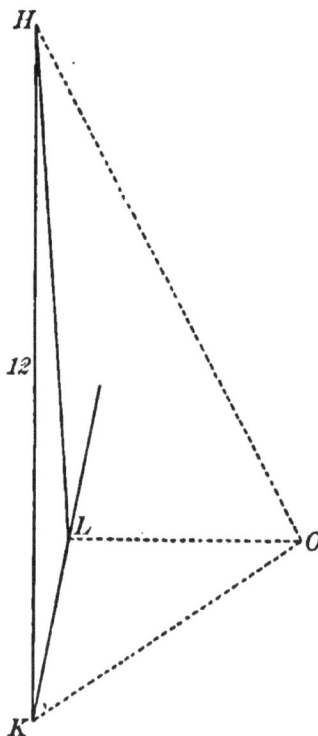

FIG. 99.   FIG. 99 a.

Having constructed the space diagram to scale, we
see that, if we proceed as in the first solution of the
preceding example, the line *hk* meets *kl* at a point so
far removed that the figure becomes unmanageable.
The other solution, however, is applicable, and furnishes
the neatest solution of the problem, thus exhibiting the

practical use of the *third method* described in Art. 80. This the student should work out for himself.

Otherwise.—Through the middle point of AB draw the straight line hk vertically downwards, and draw a straight line HK vertically downwards, and of length 12 units, to represent the weight of the rod.

Mark the straight line BC with the letters kl, and draw KL parallel to kl, the position of the point L being at present unknown.

Mark AC with the letters ol, and through A and C draw straight lines oh and ok respectively, to meet at some point on the line hk.

Let the straight lines through H and K parallel to ho and ko respectively meet in O. Draw OL parallel to ol to meet KL in L.

Then KL represents the tension of the string, and LH represents the action of the hinge upon the rod at A.

We find that $KL = 3·06$ units and $LH = 9·02$ units; also the angle LHK is found to be $4°$. Hence the tension of the string is $3·06$ pounds' weight, and the action at the hinge is $9·02$ pounds' weight in a direction inclined at an angle of $86°$ to the horizontal.

86. Ex. 3. *Two thin straight rods AFB, CFD, of given lengths, are freely jointed together at F, the lengths BF, DF being given; and the whole is laid on a smooth horizontal table. If A and C are connected by a fine string of given length, and B and D are pulled apart by given forces P, P in the straight line BD, required to find the tension of the string.*

The data are sufficient to enable us to construct the space diagram. Let T be the measure of the tension of the string.

Consider the forces acting on the rod AFB alone. It is kept in equilibrium by a given force P acting along DB, a force T of unknown magnitude along AC, and the reaction R of the hinge at F of unknown magnitude and in an unknown direction. Also B, C, F are points in the lines of action of these three forces respectively.

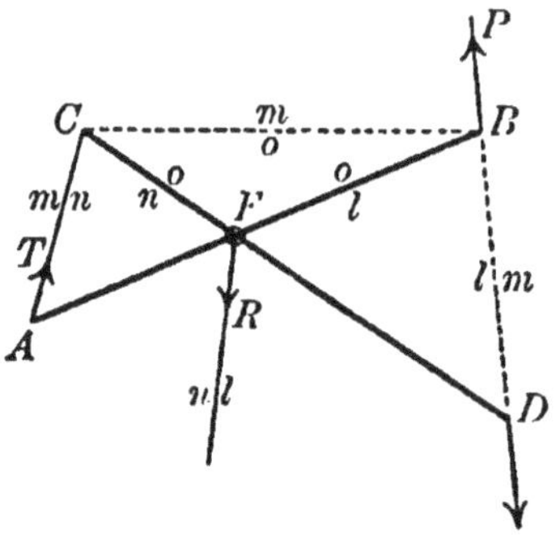

FIG. 100. 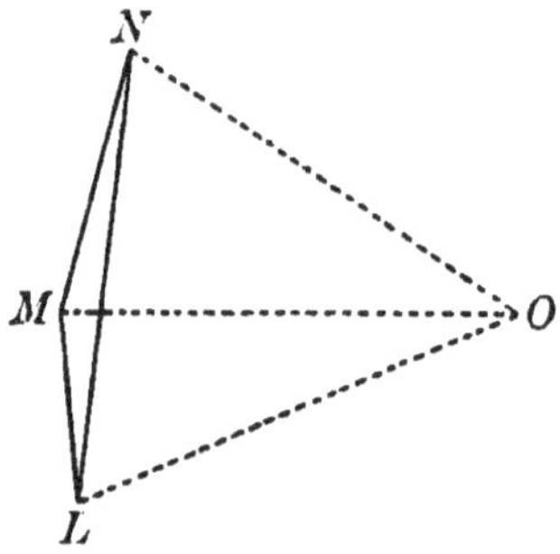   FIG. 100 a.

Hence, to construct the force diagram, with any suitable scale we draw LM of length P units in the direction DB, and MN of unlimited length in direction AC, so that the point N is not yet determined. We now mark the lines DB, AC with the letters lm, mn respectively. The line of action of R, at present unknown, but passing through F, will be called nl. Hence, also, the lines FB, BC, CF we now mark ol, om, on respectively.

Through L and M draw LO, MO parallel to lo, mo respectively. This gives us the point O. Draw ON parallel to on, to meet MN in N; then, joining NL, we complete the force diagram. $LMNL$ (this way round) is the triangle of forces for the rod AB. Measuring MN, we have the tension of the string.

87. Ex. 4. *A thin rod, of no appreciable weight, is loaded at some point with a mass of given weight, and is supported horizontally upon two smooth pegs in given positions. It is required to determine the pressures between the rod and the pegs for different positions of the load.*

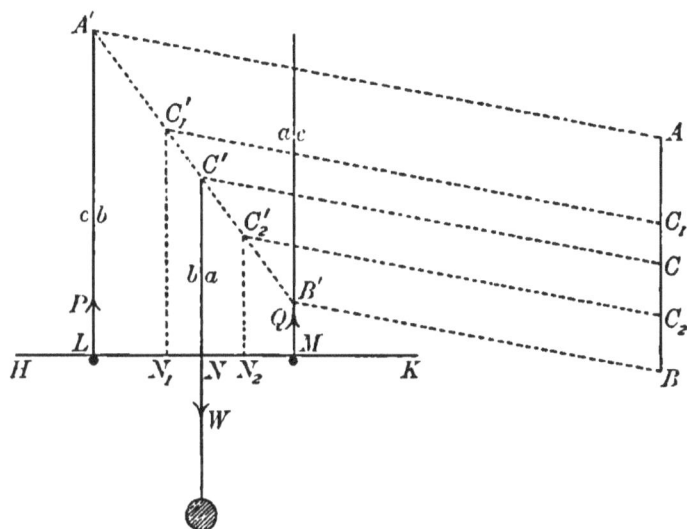

FIG. 101.

Let *HK* represent the rod, resting upon the pegs at *L* and *M*, and loaded at *N* with a mass of weight *W*. Let *P* and *Q* be the measures of the pressures of the pegs at *L* and *M* respectively upon the rod. These must be vertical, and are at present of unknown magnitude.

Let the verticals through *L*, *M*, *N* be called *bc*, *ca*, *ab* respectively. Then the force diagram will be a straight line *ACB*, in which *AB* is vertically downwards and of length *W* units, and *BC*, *CA* represent the pressures *P* and *Q* respectively. Also, if three parallels through

A, B, C meet bc, ca, ab in A', B', C' respectively, the points A', B', C' must be collinear.

I. Suppose the position of N is known. We can draw AB and two parallels AA', BB'. Let $A'B'$ meet ab in C'. Then draw $C'C$ parallel to $A'A$, to meet AB in C. Measure BC and CA, and we have P and Q respectively.

II. Suppose that the pegs L and M cannot sustain pressures greater than P_0 and Q_0 respectively. We can draw AB as before and the two parallels AA', BB'. Take BC_1 along BA and AC_2 along AB of lengths P_0 and Q_0 units respectively. If BC_1 and AC_2 do not overlap, clearly it will be impossible to support the load. If they do overlap, the point C may lie anywhere between C_1 and C_2. Draw C_1C_1', C_2C_2' parallel to AA', to meet $A'B'$ in C_1' and C_2' respectively. Then C' must lie between C_1' and C_2'. Draw $C_1'N_1$ and $C_2'N_2$ perpendiculars upon HK. Then the point of attachment of the load may lie anywhere between N_1 and N_2.

If it is required to place the load so that P may be as much less than P_0 as Q is less than Q_0, then C must be taken midway between C_1 and C_2. Hence C' will be the middle point of $C_1'C_2'$, and N of N_1N_2.

88. Ex. 5. *Three forces, represented by $B'C'$, $C'A'$, $A'B'$, act at points A, B, C respectively of a rigid body, and are in equilibrium. The line of action of the first force meets BC in a; $A'a'$ is drawn parallel to BC to meet $B'C'$ in a'. Prove that a' divides $B'C'$ in the same ratio that a divides BC.*

Since $B'C'$, $C'A'$, $A'B'$ represent forces in equilibrium, acting at A, B, C respectively, therefore, by Art. 77, the lines through A', B', C', parallel to BC, CA, AB

respectively, are concurrent in some point O. Thus, the straight lines through C' and B', parallel to AB, AC respectively, meet at a point O on $A'a'$.

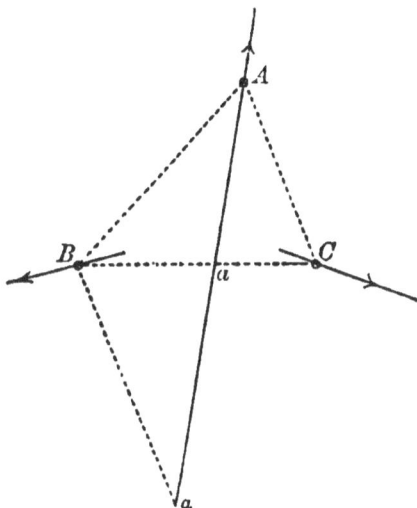 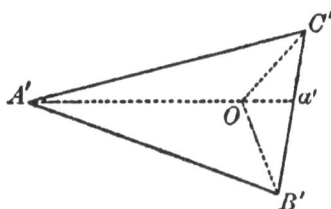

FIG. 102.   FIG. 102a.

Through B draw a straight line parallel to AC, to meet Aa in a. Then the figure $BaaA$ is similar to the figure $OB'a'C'$.

∴ a divides aA in the same ratio that a' divides $B'C'$. But the lines BC, aA are similarly divided at a.

∴ a' divides $B'C'$ in the same ratio that a divides BC.

89. Ex. 6. *A rigid body is capable of turning freely in one plane about a fixed point H, and is acted upon by a given force, represented by AB and acting along a given straight line ab. An unknown force X is applied at a given point K of the body. It is required to represent the different values, corresponding to differ- ent directions, of X consistent with equilibrium.*

The force of constraint at H may be of any value and in any direction.

Let the line HK be marked oc, and draw through H and K straight lines oa, ob respectively, to meet at any point chosen on the line ab.

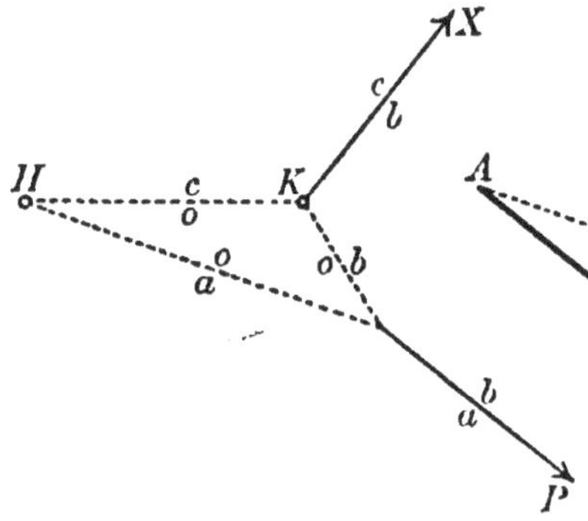

FIG. 103. 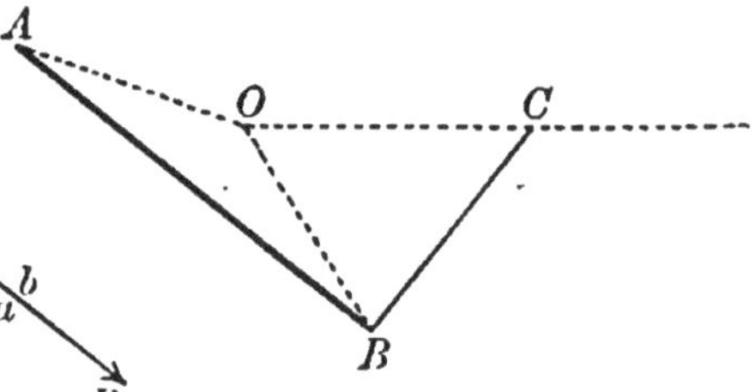   FIG. 103 a.

Through A and B draw straight lines parallel to ao, bo respectively, meeting in O; and through O draw OC parallel to oc.

Then the different values of X are represented by the different straight lines drawn from B to different points C in OC.

The shortest of the lines BC will be the perpendicular from B on OC. Hence X is smallest when it is applied in a direction perpendicular to HK.

90. Ex. 7. *A uniform thin rod HK, of given weight w, is capable of turning freely in a vertical plane about the point H, which is fixed. A mass, of given weight W, hangs from the point K by means of a fine light string. Show how to find the magnitude of the force which must be applied at K in a given direction, in order that the rod may rest in a given position.*

In the position of equilibrium, the string hangs with
the mass vertically below K. Draw AB, BC vertically
downwards to represent w and W respectively, and bisect
AB in O. Then AC represents the resultant of w and
W, and AO, OB represent components into which this
resultant can be resolved acting at H and K respectively.

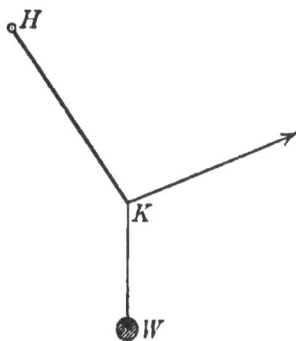

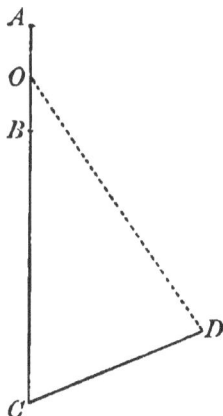

FIG. 104.   FIG. 104 a.

Draw OD parallel to HK, to meet in K the straight
line drawn through C parallel to the given direction
of the applied force. Then, measuring CD, we have
the magnitude of the applied force.

91. Ex. 8. *A heavy thin rod of given weight, whose
centre of gravity is in a given position, rests on two
given smooth inclined planes whose intersection is a
horizontal line, the rod lying in a vertical plane perpendicular to this line of intersection. It is required
to find the direction of the rod, in the position of
equilibrium, and the pressures on the planes.*

Let the vertical plane containing the rod be the plane
of the paper, cutting the inclined planes in the lines
OA, OB.

At the outset we cannot indicate correctly the position of the rod. Suppose that CD represents the rod in the position of equilibrium, and let G be its centre of gravity. Then the lengths CG, GD are known, but not the positions of C and D.

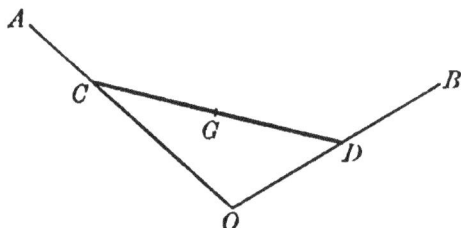 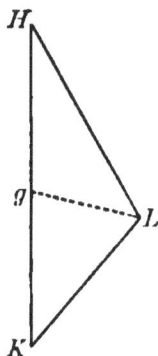

FIG. 105.   FIG. 105 a.

The rod is in equilibrium under the influence of its weight and of the reactions at C and D. The weight is equivalent to a single force acting vertically downwards through G, and, as the planes are smooth, the reactions at C and D are perpendicular to OA and OB respectively.

Draw HK vertically downwards to represent the weight of the rod, and through H and K draw straight lines perpendicular to OB and OA respectively, to meet in L. This can be done although we do not know the positions of C and D. Then measuring KL, LH we have the pressures of the planes upon the rod.

Divide HK in g so that $Hg : gK = CG : GD$. Then the weight of the rod is equivalent to two forces represented by Hg, gK acting at D and C respectively. Replacing the weight of the rod by these components, we see that the forces represented by LH and Hg, acting at

D, balance the forces represented by gK and KL, acting at C.

∴ gL is parallel to CD.

Hence, to determine the direction of the rod in the position of equilibrium, we find the points L and g as above; then the rod rests in a position parallel to gL. (For a complete discussion of this problem, see Minchin's *Statics*.)

EXAMPLES IX.

1. A straight uniform thin rod AB of mass 12 pounds, capable of turning freely in a vertical plane about a fixed point A, rests in a horizontal position with its extremity B in contact with a smooth plane inclined at an angle of 30° to the horizon. Find the actions at A and B.

2. A uniform thin rod, 12 feet long and weighing 50 pounds, is capable of turning freely about its lower end A, and a point C of the rod, distant 10 feet from A, is connected by a horizontal fine string CD to a point D, situated 8 feet vertically above A. Find the tension of the string and the reaction at A.

3. A straight thin rod AD, of no appreciable weight and of length 3 feet, is capable of turning freely in a vertical plane about its lower extremity A. The point C of the rod, distant 2 feet from A, is connected by a light inextensible string 1 foot 6 inches long to a point B, fixed 2 feet 6 inches vertically above A. At D is attached another fine string supporting a mass of 100 pounds. Find the tension of the string.

4. A straight thin rod AB, of no appreciable weight and of length 24 inches, is capable of turning freely in a vertical plane about the extremity A, which is fixed. A fine string, of length 18 inches, has one end attached to a point C of the rod distant 15 inches from A, and the other end to a fixed point D, situated 20 inches vertically above A. A mass of 100 pounds is suspended from the point B. Find the tension of the string.

5. A uniform beam, 12 feet in length, has a fixed hinge at one end, and is supported by a fine light cord, 13 feet long, attached to the other end and to a fixed point situated 20 feet vertically above the hinge. Find the tension of the cord, assuming that the beam weighs 140 pounds.

6. A uniform straight rod AOB, of mass 30 pounds, is capable of turning freely about a hinge at O,—a point dividing AB in the ratio $1:3$. The rod rests with its lower end B in contact with a smooth inclined plane. If rod and plane are each inclined at an angle of 30° to the horizon, and the line of greatest slope of the plane through B is in the same vertical plane as the rod, find the pressure at B and the reaction of the hinge.

7. A straight rod AOB, of no appreciable weight, is capable of turning freely in a vertical plane about a fixed point O, such that $AO=2 \cdot OB$. A mass of 10 pounds is suspended from B, and the rod is supported in a horizontal position by a fine string AC, connecting A with a point C situated vertically below O. If the string makes an angle of 30° with the horizon, find its tension and the action at O.

8. A uniform straight rod AB, of weight W, rests in an inclined position with the end B against a smooth vertical wall, and the end A is fixed in position by a smooth hinge. If the height of B above A is to the horizontal distance between B and A as $3:8$, express the forces which keep the rod at rest in terms of W.

9. A uniform horizontal beam AB, of length 12 feet and weighing 100 pounds, is placed with the end A against a rough vertical wall AD, and is supported by a fine string CD, of length 10 feet, connecting the point C of the beam, distant 8 feet from A, with a point D in the wall, situated vertically above A. Find the tension of the string, and the resistance of the wall. If the beam is just on the point of slipping, determine the coefficient of friction between the beam and the wall.

10. A uniform rod AB, of length 2 feet and weighing 55 ounces, rests with its lower end A in contact with a smooth vertical wall, being inclined to the wall at an angle of 40°. It is supported by a fine string connecting the point C of the rod, distant 8 inches

from A, with a point D in the wall, situated vertically above A. Find the length of the string and its tension.

11. A fine rod BC, of no appreciable weight and of length 2 feet 6 inches, is capable of turning freely about the extremity B, which is fixed. A load of 150 pounds is applied at C, and the rod BC is supported in a horizontal position by means of another fine light rod DE, of length 1 foot 10 inches, smoothly hinged at E to the rod BC and at D to a fixed point vertically below B. If BE is of length 1 foot 6 inches, find the action at B and the thrust in DE.

12. A rod ACB, weighing 25 ounces, rests upon a smooth peg C, and its end A is attached to a fixed point O, in the same horizontal line with C, by means of a fine string OA. If $OA = OC = 1$ foot, and the rod rests at an angle of 25° to OC, determine the position of the centre of gravity of the rod, and the magnitudes of the tension of the string and the pressure between the rod and the peg.

13. A square lamina $ABCD$, of uniform density and weighing 4 pounds, can turn in a vertical plane about a hinge at A. Find the force which, acting along BC, will keep the lamina in a position with this side horizontal and below AD; find also the magnitude and direction of the hinge action at A.

14. A uniform square lamina $ABCD$ is capable of turning in a vertical plane about a smooth hinge A. It is kept in equilibrium with AB inclined to the horizon at an angle of 30°, measured downwards, by a horizontal force of 2 pounds' weight applied at B. Find the mass of the lamina.

15. A uniform square lamina $ABCD$, of mass 4 pounds, is capable of turning freely about the point A, which is fixed; a fine string, of length equal to a side of the square, connects B with a point E, situated in a horizontal line with A. Find the tension of the string, the angle BAE being 20°.

16. A rectangular box, containing a uniform spherical ball of weight W, stands on a horizontal table, and is tilted about one of its lower edges through an angle of 30°. Find the pressures between the ball and the box.

17. A heavy uniform sphere, of weight W, rests on a smooth plane inclined at an angle of 60° to the horizon. It is supported by a fine string of length equal to the radius of the sphere, connecting a point in the surface of the sphere with a point in the surface of the plane. Find the tension of the string in the position of equilibrium, and the pressure between the sphere and the plane.

18. A uniform spherical ball, whose radius is 1 foot and mass 8 pounds, is fastened by a fine string, 8 inches long, attached to its surface and to a smooth vertical wall. Find the pressure on the wall and the tension of the string.

19. A smooth uniform sphere, of mass 60 pounds and diameter 10 inches, is supported in contact with a smooth vertical wall by a fine string, 8 inches long, fastened to a point on its surface, the other end being attached to a point in the wall. Find the tension of the string.

20. A homogeneous solid sphere, of diameter 10 inches and weighing 30 pounds, rests upon a smooth inclined plane, whose height is $\frac{4}{5}$ of its length, being supported by a fine string, 8 inches long, connecting a point in the surface of the sphere with a point on the plane. Find the tension of the string, and the pressure between the sphere and the plane, in the position of equilibrium.

21. A smooth uniform sphere, of mass 52 pounds and radius 10 inches, rests on a smooth inclined plane, whose height is $\frac{5}{13}$ of its length, against a smooth horizontal rail fixed parallel to the plane. If the pressure between the sphere and the plane is 33 pounds' weight, find the distance of the rail from the plane and the pressure between the sphere and the rail.

22. A triangular lamina ABC, of inappreciable weight, rests in a vertical plane with the middle points of the sides AB, AC in contact with two smooth pegs, the line joining them being horizontal and parallel to the base BC. Determine the point in BC where a mass of weight W may be placed without disturbing the equilibrium; and, if AB, AC, and BC be 4, 5, and 6 feet respectively, find the pressures on the pegs in terms of W.

23. A rigid framework, of no appreciable weight, in the shape
of an equilateral triangle ABC, rests in a vertical plane with
BC horizontal and uppermost. It is supported in this position
by a fine string DG, parallel to AB, attached at D, the middle
point of BC, and rests against a small smooth peg at F, the
middle point of AB. Determine from what point of the boundary
of the framework a mass of 10 pounds may be suspended, and
find the tension of the string and the pressure at F when the
mass is attached.

24. A uniform beam AB, of mass 20 pounds and length 13
inches, is capable of turning freely about a fixed point A ; to
the other end B is attached a fine string, which passes over a
small smooth pulley C, situated 2 feet in a horizontal line from A.
Find what mass must be attached to the other end of the string,
in order that, in the position of equilibrium, the beam and the
string may be equally inclined to the horizontal. Find also the
action at the hinge.

25. A lamina, in the shape of a regular hexagon $ABCDEF$,
lies on a smooth horizontal table. It is in equilibrium under
the action of three forces ; namely, a force of 20 pounds' weight
acting at A in the direction EA, a force of unknown magnitude
acting at F in direction BF, and a force unknown both in magnitude and direction acting at C. Determine the magnitudes of
the unknown forces.

26. In example 8, Art. 91, the planes AO and OB are inclined
at angles of 60° and 30° respectively to the horizontal. The point
G divides CD in the ratio 3 : 1. Find the direction of the rod in
the position of equilibrium.

27. $OACB$ is a horizontal straight line. C is the centre of a
fixed vertical circular disk ADB. A uniform rod OD, of length
equal to the radius of the circle and weighing 2 pounds, is freely
hinged at one extremity to a fixed point O, and its other extremity
D rests against the smooth rim of the circle. If the rod makes
an angle of 30° with OB, find all the forces which act upon it.

28. A smooth rod BC is passed through a small ring and placed
upon a horizontal plane, with its ends attached to a fixed point
A in the plane by two fine strings AB, AC, which are tight.

A horizontal force being applied to the ring, find its direction, and also the position of the ring on the rod, in order that equilibrium may not be disturbed, the lengths of BC, CA, AB being 25, 20, and 15 inches respectively.

29. A smooth thin wire APB, in the form of a semicircle of radius 15 inches, is placed upon a smooth horizontal table with its ends attached to a fixed point O by means of fine strings AO, BO, which are tight. A small ring P is threaded on the wire, and a horizontal force is applied to the ring; find its direction, and also the position of the ring on the wire, in order that equilibrium may not be disturbed, the lengths of AO, OB being 24 and 18 inches respectively.

30. A straight piece of stiff wire AB, 13 inches long and of no appreciable weight, is capable of turning freely in a vertical plane about one extremity A. A small smooth ring, of no appreciable weight, is threaded on the wire and connected by a fine string, 12 inches long, to a point fixed 13 inches vertically above A. A mass of $6\frac{1}{2}$ pounds is hung from B. Find the tension of the string, and the action at the hinge, in the position of equilibrium.

31. A uniform rod, of weight W, is supported by a fine string fastened to its ends, of double its own length, which passes over a smooth horizontal rail. Find the tension of the string, first, when the rod is hanging at rest in a vertical position, and secondly, when the rod is at rest in a horizontal position.

32. A straight thin rod AB, of length 1 foot and of no appreciable weight, is supported in a horizontal position upon two pegs situated at its extremities. If the peg at A cannot sustain a load greater than 27 pounds, and the peg at B cannot sustain a load greater than 24 pounds, find between what points of the rod a load of 36 pounds may be placed.

33. In the preceding example, find where the load must be placed in order that the pressure at A may be 21 pounds' weight.

34. A uniform rod, 2 feet long and weighing 3 pounds, lies on a horizontal plane ; find the least force which, applied 5 inches from one end, will raise that end above the plane.

35. A horizontal bar AB, 7 feet long, is supported at its extremities, and a man of 150 pounds' weight hangs from it by his hands, one being 1 foot from A, the other 3 feet from B. Find the pressures on the supports due to the weight of the man.

36. A heavy pole, weighing 140 pounds, is carried on the shoulders of two men, one at each end; the centre of gravity of the pole being 2 feet from one end and 5 feet from the other, find the load supported by each man.

Also, find what would be the effect of placing each man one foot nearer to the centre of gravity of the pole.

37. Two levers AOB, COD, of inappreciable weight, whose lengths are 8 and 9 inches respectively, are freely jointed together at O, four inches from B and D. If A and C are connected by a fine light string 3 inches long, and B and D are pulled apart by forces each equal to 10 pounds' weight in the straight line BD, find the tension of the string.

38. Two levers OA, OB, of inappreciable weight and of lengths 3 and 4 feet respectively, can turn freely in a vertical plane about a common fulcrum O, and their middle points are connected by a fine string whose length is $2\frac{1}{2}$ feet. Find the least force which, applied at A, will keep OB horizontal when a mass of 12 pounds is suspended from B. Find also the tension of the string.

39. A uniform thin rod AB, which can turn freely in a vertical plane about a hinge at A, is kept in a horizontal position by a string BC attached to a fixed point C in the vertical plane, the angle ABC being obtuse. Show in a diagram the forces acting on the rod, and prove that two of them are equal.

40. If a heavy body is partly supported by a string and partly by a smooth horizontal plane, prove that the string must be vertical.

41. Three forces P, Q, R, in equilibrium, act at points A, B, C respectively of a rigid body. The lines of action of two of the forces meet at O, and the circle described through the points B, C, O meets AO again in A'. Prove that $P : Q : R = BC : CA' : A'B$.

42. Three forces P, Q, R, in equilibrium, act at points A, B, C respectively of a rigid body. Prove that if two of the forces

intersect at any point on the circle which circumscribes the triangle ABC, then $P : Q : R = BC : CA : AB$.

43. Three forces act at points A, B, C of a rigid body, and are respectively proportional to BC, CA, AB. Prove that, if the forces are in equilibrium, their lines of action must intersect *either* at the orthocentre of the triangle ABC, *or* at a point on the circle which circumscribes the triangle.

44. Three forces, whose magnitudes are in a given ratio, act at points A, B, C respectively of a rigid body. Show how to determine the lines of action of the forces, in order that they may be in equilibrium. Show that there are in general two solutions.

45. If, in the preceding example, O_1 and O_2 are the two positions of the point at which the lines of action of the three forces intersect, prove that each side of the triangle ABC subtends at O_1 the same angle that it subtends at O_2.

46. Two forces Q and R, acting in the lines OB, OC respectively, are in equilibrium with a third force, acting through A. Prove that, if $Q : R = CA : AB$, then, *either* O is a point on the circle which circumscribes the triangle ABC, *or*, if BO and CO meet that circle in O_1 and O_2 respectively, the arc O_1O_2 is bisected at A.

47. A given force, acting along a given straight line Aa, is in equilibrium with two unknown forces acting through given points B and C respectively. If Aa is perpendicular to BC, prove that the difference between the squares of the measures of the unknown forces is constant.

48. B, A, C are three given points situated in a straight line. A force of given magnitude, acting through A in a given direction, is in equilibrium with two unknown forces, whose magnitudes are in a given ratio, acting through B and C respectively. Show how to determine the magnitudes and directions of the unknown forces, and prove that there are in general two solutions.

49. In the preceding example, prove that if the forces through B and C are proportional to CA, AB respectively, then, *either* all the forces are parallel, *or* the forces through B and C are equally inclined to BC.

50. A rigid body, capable of turning freely about a fixed point c, is kept in equilibrium by two forces P and Q represented by BC, CA respectively. A straight line through c meets the lines of action of P and Q in a and b respectively; and a straight line through A parallel to bc meets BC in A'. Prove that A' divides BC in the same ratio that a divides bc. In particular, if cab is drawn making equal angles with P and Q, prove that $ca : cb = Q : P$. Hence prove that the perpendiculars from c on the lines of action of P and Q are proportional to Q and P respectively.

51. Three forces in equilibrium act along lines which bisect at right angles the sides of a triangle. Prove that the forces must act all outwards or all inwards, and that their magnitudes are proportional to the sides to which they are respectively perpendicular.

CHAPTER X.

RESULTANT OF ANY SYSTEM OF COPLANAR FORCES.

92. *To find the resultant of any given system of forces acting upon a rigid body in one plane.*

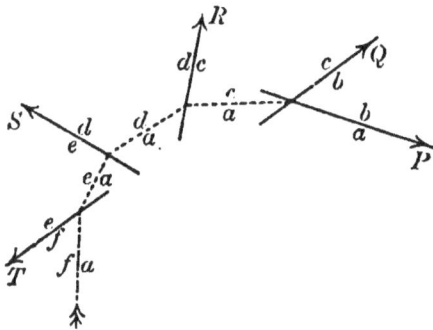

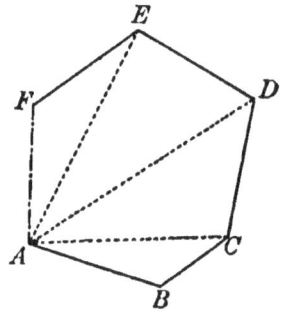

FIG. 106.   FIG. 106 a.

First Method.

Let it be required to find the resultant of forces *P*, *Q*, *R*, *S*, *T*, acting along the given lines indicated in the diagram on the left.

Draw straight lines *AB*, *BC*, *CD*, *DE*, *EF*, to represent the forces *P*, *Q*, *R*, *S*, *T* respectively in direction and magnitude. This takes us from *A* to *F*. We shall show that *AF* represents the resultant in direction and magnitude, and that its line of action may also be found.

Let the lines of action of P, Q, R, S, T be now marked ab, bc, cd, de, ef respectively, as in the figure. Join AC, AD, AE, AF.

The resultant of P and Q is represented by AC, and acts through the point of intersection of ab and bc; hence draw a straight line ac through this point parallel to AC. Let P and Q be replaced by their resultant, which we can combine with R. We get as the resultant of *these* two forces a force represented by AD, passing through the point of intersection of ac and cd. Hence P, Q, R are together equivalent to a force represented by AD, and acting along a line ad, which we can draw through the point of intersection of ac and cd parallel to AD.

Proceeding in the same way as before, we see that the forces P, Q, R, S may be replaced by a force represented by AE, and acting along a line ae, which we can draw through the point of intersection of ad and de parallel to AE. And, finally, the forces P, Q, R, S, T are equivalent to a force represented by AF, and acting along a line af, which we can draw through the intersection of ae and ef parallel to AF.

This determines the magnitude, direction, and position of the resultant of the system.

93. In order that the system may be in equilibrium, it will be necessary and sufficient that the force T should be equal and opposite to, and act along the same straight line as, the resultant of P, Q, R, S. For this to be the case, F must coincide with A, and the two straight lines ae and fe must form one continuous straight line. That is, F must coincide with A, and the straight line ad, obtained by the above process, must pass through the intersection of *de* and *ef*. Thus, for equilibrium, in constructing the force diagram, the end of the line representing the last force must coincide with the beginning of the line representing the first force (this is expressed by saying *the force polygon closes*), and in drawing the dotted lines of the space diagram, we start with the point of intersection of the first two lines of action, and end with the point of intersection of the last two.

94. The method of Art. 92 may be still more abbreviated in simple cases, as, for instance, in the following:

It is required to find the resultant of four given forces, P, Q, R, S, acting along the given lines indicated in the space diagram.

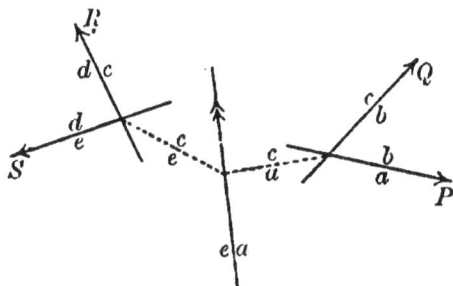 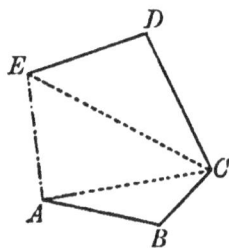

FIG. 107.   FIG. 107 a.

Draw straight lines *AB*, *BC*, *CD*, *DE*, to represent the forces *P*, *Q*, *R*, *S* respectively in magnitude and direction. Let the lines of action of *P*, *Q*, *R*, *S* be now marked *ab*, *bc*, *cd*, *de* respectively, as in the figure. Join *AC, AE, CE*. Through the intersection of *ab* and *bc* draw *ac* parallel to *AC*; and through the intersection of *cd* and *de* draw *ce* parallel to *CE*. Then we shall show that the straight line *ae*, drawn through the intersection of *ac* and *ce* parallel to *AE*, is the line of action of the resultant, and that AE represents the resultant in magnitude and direction.

For, the forces P and Q are equivalent to a force represented by AC acting along ac; also, the forces R and S are equivalent to a force represented by CE acting along ce; and, if P, Q, R, S be replaced by these two forces, they in turn may be replaced by a force represented by AE, and acting along ae.

For the system to be in equilibrium, it is necessary and sufficient that A and E should coincide, and that the lines ac, ce should form one continuous straight line.

Thus, for equilibrium, the force diagram must be a closed quadrilateral, and a diagonal of the quadrilateral must be parallel to the line joining the intersection of one pair of lines of action with the intersection of the other pair.

95. *Second Method.*

Let it be required to find the resultant of forces P, Q, R, S, T, acting along the given lines indicated in Fig. 108.

As before, draw straight lines AB, BC, CD, DE, EF, to represent the forces P, Q, R, S, T respectively in magnitude and direction, and let the lines of action of P, Q, R, S, T be now marked ab, bc, cd, de, ef respectively.

Take any point O in the force diagram, and join it to the points A, B, C, D, E, F.

From any point p in ab draw straight lines oa, ob parallel to OA, OB respectively. The force P may be replaced by two forces represented by AO, OB acting along the lines ao, ob respectively.

From the point of intersection of ob and bc draw oc parallel to OC. Then the force Q may be replaced by two forces represented by BO, OC acting along bo, oc respectively.

Let the forces P and Q be replaced by these pairs of components; then we may remove the two forces represented by OB and BO, which act in opposite directions along the line ob. Thus the forces P and Q are equivalent to forces represented by AO, OC acting along ao, oc respectively.

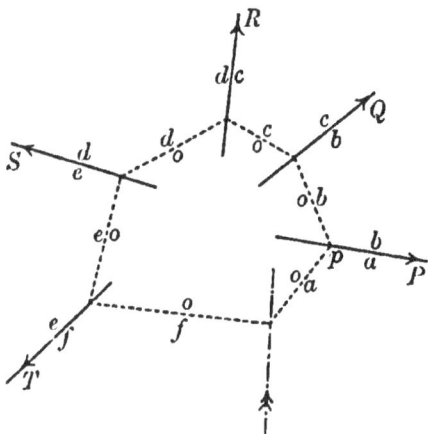 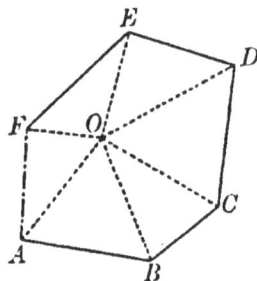

FIG. 108.   FIG. 108 a.

From the point of intersection of oc and cd draw od parallel to OD. Then the force R may be replaced by two forces represented by CO, OD acting along co, od respectively. So, removing the two equal forces which act in opposite directions along the line oc, we see that the forces P, Q, R are equivalent to forces represented by AO, OD acting along ao, od respectively.

Carrying on the process, we draw through the intersection of od and de a straight line oe parallel to OE; and through the intersection of oe and ef a straight line of parallel to OF. Then we see that the forces P, Q, R,

S, T are equivalent to two forces represented by AO, OF acting along ao, of respectively.

Hence the resultant of the system is a force represented by AF, and acts along a straight line af drawn through the intersection of oa and of parallel to AF.

96. The broken line $ABCDEF$ is called a *force polygon*, and is said to *close* when the final point F coincides with the initial point A.

The polygon formed by the lines oa, ob, oc, od, oe, of is called a *funicular polygon*, and is said to *close* when the lines oa, of are in one and the same straight line.

The point O is called the *pole* of the force polygon. The lines drawn from O to the angular points of the force polygon are called *rays*. The sides of the *funicular polygon* are called *strings*.

It should be noticed that the two strings which intersect in the line of action of any force are parallel to the two rays drawn to the extremities of the line representing that force, and the string which connects the lines of action of two forces is parallel to the ray drawn to the common extremity of the lines representing those forces.

97. In order that the system may be in equilibrium, it will be necessary and sufficient that the two forces represented by AO, OF, and acting along the lines ao, of respectively, should be equal and opposite, and act along the same straight line. For this to be the case, F must coincide with A, and of, oa must be in one and the same straight line. Thus, for equilibrium, *a force polygon must close, and a funicular polygon corresponding to it must also close.*

If the force polygon closes, but not the funicular

polygon, the lines *oa*, *of* become parallel, and in this case the system reduces to a couple.

98. Of the two methods given above, the second is the more general and includes the first; for, if we take the pole *O* to coincide with *A*, and take the first vertex of the funicular polygon to coincide with the point of intersection of the lines of action of *P* and *Q*, the construction coincides with that given in the first method.

In simple cases the first method may be more suitable than the second, but it is liable to fail through the intersections of lines falling at inconvenient distances. In the second method the pole can generally be so chosen that none of the rays make very acute angles with the corresponding forces.

99. In the case of parallel forces, the first method fails altogether, but the second method does not.

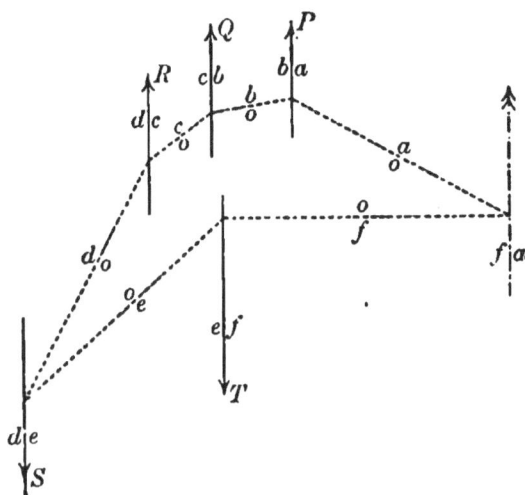 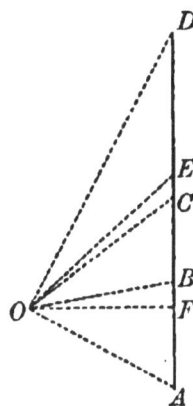

Fig. 109.   Fig. 109 a.

We append the figures for finding the resultant of the five parallel forces *P*, *Q*, *R*, *S*, *T*, using the same letters as above. The forces S and T are taken to be in the opposite direction to the other three.

The force polygon for a system of parallel forces reduces to a straight line, which is often called *the line of loads*.

100. Some Geometrical Properties of the Funicular Polygon.

Referring to Art. 95, we showed that the forces P, Q, R are equivalent to forces represented by AO, OD acting along ao, od respectively.

∴ the point of intersection of ao, od lies on the line of action of the resultant of P, Q, R, which is parallel to AD.

Now, in constructing the funicular polygon, we took O in any arbitrary position in the force diagram, and the point p was taken in any arbitrary position on the line ab. If we vary the position of p without altering the position of O, we get a new funicular polygon, having its sides parallel to the corresponding sides of the old funicular polygon. If we vary the position of O, we get a funicular polygon with its sides no longer parallel to their former directions. But, whatever alterations we make in these respects, the lines oa, od always intersect on a fixed straight line parallel to AD. Thus,

If different funicular polygons be constructed for the same system of forces corresponding to the same force polygon, the locus of the intersection of any two strings is a straight line parallel to the line joining the extremities of the corresponding rays.

101. Again, suppose two funicular polygons are constructed, one corresponding to a pole O, and the other to a pole O'.

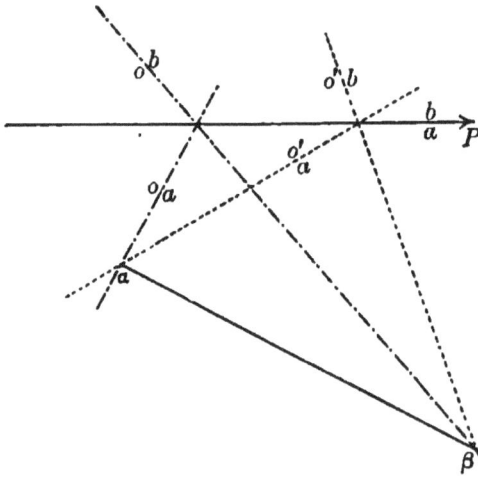
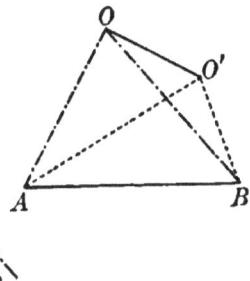

Fig. 110.   Fig. 110 a.

Let oa, $o'a$ meet in a, and ob, $o'b$ in β. We shall show that $a\beta$ is parallel to OO'.

Suppose that forces represented by AO, OB, BO', $O'A$ were to act along the lines oa, ob, $o'b$, $o'a$ respectively. They would be in equilibrium; for the first two forces are equivalent to a force represented by AB acting along ab, and the other two to a force represented by BA acting along the same straight line.

$\therefore$ forces represented by OB, BO', acting along ob, $o'b$ respectively, are in equilibrium with forces represented by $O'A$, AO acting along $o'a$, oa respectively. The first two of these forces are equivalent to a force represented by OO' acting through β, and the other two to a force represented by $O'O$ acting through a.

$\therefore$ a force represented by OO' acting through β is in equilibrium with an equal and opposite force acting through a.

$\therefore a\beta$ is parallel to OO'.

In the same way, if oc and $o'c$ meet at γ, we can show that $\beta\gamma$ is parallel to OO'.

$\therefore$ the points a, β, γ lie on a straight line parallel to OO'.

Proceeding in this way, we see that

Each pair of corresponding sides of any two funicular polygons of a given system of forces intersect on a straight line, which is parallel to that joining the poles of the two funicular polygons.

102. Ex. *Three forces P, Q, R act along three fixed lines AB, BC, CD respectively. Prove that, if P, Q, R have any values subject to the relation $Q = m \cdot P + n \cdot R$, where m and n are any given numbers or fractions, then the line of action of the resultant of the three forces passes through a fixed point.*

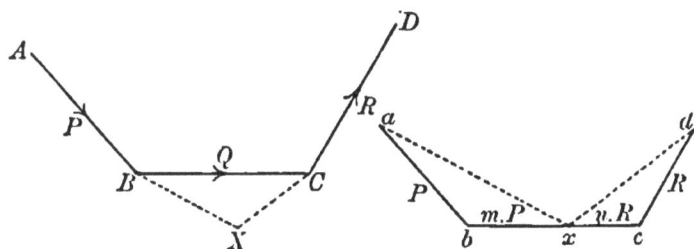

FIG. 111. FIG. 111 a.

Let ab, bc, cd be drawn parallel to AB, BC, CD respectively, representing P, Q, R respectively. Since $Q = m \cdot P + n \cdot R$, we can find a point x in bc such that $bx = m \cdot ab$ and $xc = n \cdot cd$.

Through B and C draw straight lines parallel to ax

and dx respectively, to meet in X. Then we shall show that X is a fixed point on the line of action of the resultant of the forces P, Q, R.

The force Q may be replaced by forces $m.P$ and $n.R$, acting along BC at points B and C respectively. The resultant of P and mP at B is represented by ax, and therefore acts along BX; the resultant of nR and R at C is represented by xd, and therefore acts along XC. Hence X is a point in the line of action of the resultant.

Now since ab and bx are drawn in fixed directions, and such that $ab:bx=1:m$, therefore the straight line ax is in a fixed direction. Therefore BX is a fixed straight line. Similarly CX is a fixed straight line.

∴ the point X is the intersection of two fixed straight lines, and is therefore a fixed point.

Thus the line of action of the resultant of P, Q, R passes through the fixed point X.

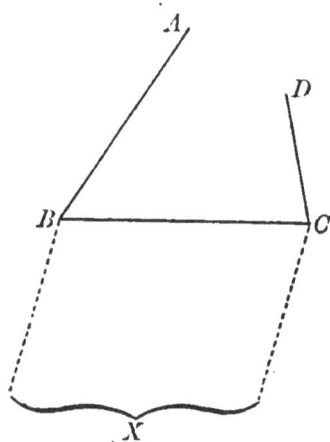

FIG. 112.

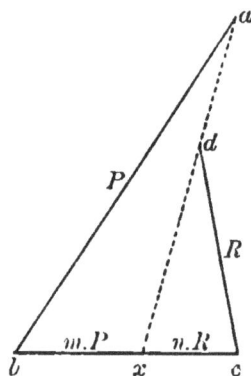

FIG. 112a.

The case in which the points a, x, d are collinear is interesting. The lines BX, CX are then parallel, and there is no point X at a finite distance. Under these circumstances the resultant of the three forces is in a fixed direction.

EXAMPLES X.

1. Forces of 2, 1, 3 pounds' weight respectively act along the sides of an equilateral triangle, taken one way round. Determine the magnitude, direction, and line of action of their resultant.

2. The sides BC, CA, AB of the triangle ABC are of lengths 14, 15, 13 inches respectively. Forces of magnitudes 16, 60, 52 pounds' weight act along the lines BC, CA, BA respectively. Find the magnitude, direction, and line of action of their resultant.

3. Forces of 1, 2, 4, 4 pounds' weight act along the sides AB, BC, CD, DA respectively of a square. Find the magnitude and direction, and the point of application in the line BC, of the force which would balance the system.

4. $ABCD$ is a square, each side of which is 1 foot in length. E is a point in AB distant 3 inches from A, F is in DA produced 14 inches from A, and G is in CB produced 9 inches from B. Find the magnitude, direction, and line of action of the resultant of the following system of forces:—45 pounds' weight along AB, 66 pounds' weight along AD, 35 pounds' weight along CE, and 65 pounds' weight along FG.

5. $ABCDEF$ is a regular hexagon. Forces of 5, 3, 5, 3 pounds' weight respectively act along the straight lines AB, BC, CD, DE. Find the magnitude, direction, and position of the resultant.

6. Take any five forces, assigning their magnitudes, directions, and lines of action, and determine the magnitude and direction of their resultant by constructing a force polygon. Determine the line of action of the resultant by constructing a funicular polygon corresponding to an arbitrarily chosen pole O.

Draw another funicular polygon corresponding to the same pole O, and a third funicular polygon corresponding to a different pole O', and see that each funicular polygon gives the same line of action of the resultant.

Construct another force polygon by taking the forces in a different order, and, taking any pole, construct a funicular polygon corresponding to this, and see that the same result is obtained as before.

7. Four forces act along, and are represented by, AB, BC, DC, AD; show that their resultant is represented by $2AC$, and acts through the middle point of BD.

8. Four forces act along, and are represented by, AB, CB, CD, AD; show that their resultant acts along, and is represented by, $4EF$, where E and F are the middle points of AC, BD respectively.

9. Three forces P, Q, R, such that $P=Q+R$, act along the sides BC, AC, BA of a triangle ABC; prove that the line of action of their resultant passes through the centre of the circle inscribed to the triangle.

10. Three forces act along the sides of a triangle, taken one way round. If one of the forces is equal to the sum of the other two, prove that the line of action of their resultant passes through the centre of one of the circles escribed to the triangle.

11. Three forces P, X, Y act along the sides BC, CA, BA of a given triangle ABC. If P is given, while X and Y have any values subject to the condition that $X+n.Y$ is constant, where n is any given number or fraction, prove that the line of action of the resultant of the three forces passes through a fixed point.

12. Three forces act along, and are represented by, AB, BC, CD; prove the following method for determining their resultant: Take any point x in BC, and let straight lines through B and C, parallel to Ax, Dx respectively, meet in X; then the resultant acts through X, and is represented by AD.

13. A, B, C, D are four fixed points; any point x is taken in BC, and straight lines through B and C, parallel to Ax, Dx respectively, meet in X. Prove, by a statical method, that as x moves along BC, the point X traces out a straight line parallel to AD.

CHAPTER XI.

EQUILIBRIUM OF FOUR FORCES HAVING KNOWN LINES OF ACTION IN ONE PLANE.

103. Before considering examples on the general case of the equilibrium of coplanar forces, we will consider the equilibrium of four forces having known lines of action situated in one plane. This we can do, in general, without drawing a funicular polygon.

Four forces in one plane are in equilibrium, and the lines of action of all are given. One of the forces is fully known; it is required to determine the other three.

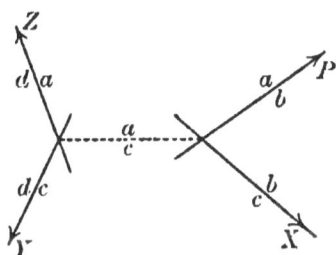 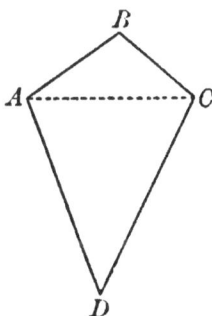

FIG. 113.   FIG. 113 a.

Let *P* be the measure of the given force acting along the given line indicated. Take *AB* in the direction of this force and make it P units of length. We now mark the line of action of the given force with the letters ab.

Find the point in which ab intersects one of the other three given lines, and let this line be denoted by bc. We then mark the remaining two lines cd, da.

Join the point of intersection of ab and bc to the point of intersection of cd and da; and let the line so drawn be called ac.

Let X, Y, Z, at present unknown, be the measures of the forces acting along bc, cd, da respectively. The resultant of P and X has to balance the resultant of Y and Z. Hence these two resultants must be equal and act in opposite directions along the same straight line. Thus ac must be the line of action of the resultant of the pair P and X, and also of the resultant of the pair Y and Z.

Hence, if BC represents the force X, then AC must be parallel to ac. So, drawing BC parallel to bc to meet in C the straight line drawn through A parallel to ac, and measuring BC, we have X. Also AC represents the resultant of P and X, and therefore CA represents the resultant of Y and Z.

Hence, to find Y and Z, we have merely to draw straight lines through C and A parallel to cd, da respectively and meeting in D; then, measuring CD and DA, we have Y and Z respectively.

If the four given lines are concurrent, the solution is indeterminate. In this case the extremities of the line ac coincide, and the line AC may be drawn in any direction.

If the three lines bc, cd, da meet in a point which

is not situated in the line of action of P, the problem is impossible of solution. In this case ac is in the same straight line with bc, and the line BC does not meet the line AC.

If two of the three given lines intersect on the line of action of P, the force along the third line vanishes. For instance, if ab, bc, cd are concurrent, ac is in the same straight line with cd, and D coincides with A.

104. If the lines cd, da do not meet at an accessible point, we can still draw the straight line ac by making use of the following construction:

To draw through a given point P a straight line towards the inaccessible point of intersection of two given straight lines AA', BB'.

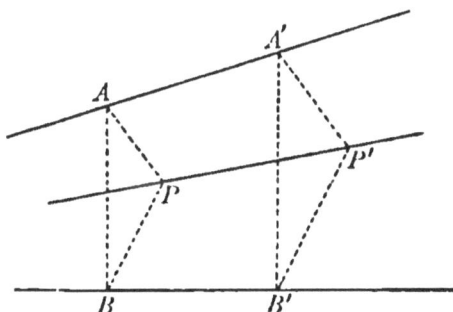

FIG. 114.

Draw straight lines PA, PB to points A and B, situated one in each of the given straight lines, and join AB. Draw a straight line parallel to AB intersecting the given straight lines in A' and B' respectively.

Draw $A'P'$, $B'P'$ parallel to AP, BP respectively, to meet in P'. Then PP' is the straight line required.

The student of elementary geometry will have no difficulty in proving the accuracy of this construction.

105. The foregoing construction enables us to apply the method of Art. 103 to the case in which ab, bc meet at an accessible point, and cd, da at an inaccessible point; but the method apparently breaks down when cd and da are parallel.

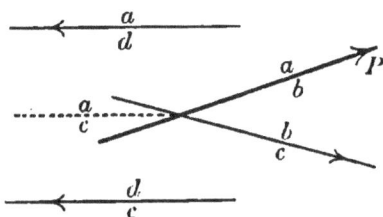 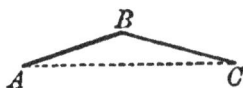

FIG. 115.   FIG. 115 a.

The difficulty, however, is easily overcome. We can draw the line AC parallel to either of the lines cd, da, thus determining the point C. Then, to finish the problem, we have merely to resolve a force represented by CA into two parallel components acting along the lines cd, da. This can be done by any one of the three methods considered in Art. 72.

106. If the line of action of the given force does not meet any one of the other three lines in an accessible point, we may proceed as follows:

As before, take AB to represent the given force, and let ab denote its line of action. The three unknown forces will be represented by BC, CD, DA, where the points C and D are to be found. Let their lines of action be therefore marked bc, cd, da.

The two forces along cd, da are equivalent to a force which will be represented by CA, and which acts through the point of intersection of cd and da, along a straight line which would be marked ca. It will not be necessary to draw the straight line ca, but we may refer to it.

Draw any straight line *ob* intersecting the lines *ab* and *bc*. From the point of intersection of *ob* and *ab* draw *oa*, and from the point of intersection of *ob* and *bc* draw *oc*, both to the point of intersection of *ad* and *cd* (that is, to a point in *ca*); this can be done even if *ad* and *cd* do not intersect at an accessible point.

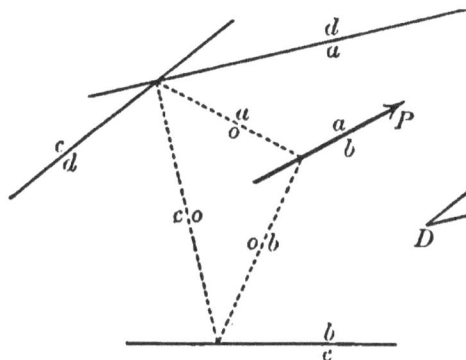
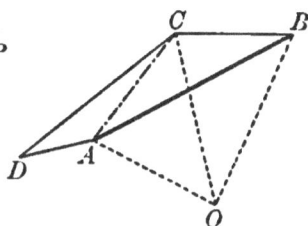

FIG. 116.   FIG. 116*a*.

Draw *AO*, *BO* parallel to *oa*, *ob* respectively, to meet in *O*. Draw *OC*, *BC* parallel to *oc*, *bc* respectively, to meet in *C*. Then, if *C* be joined to *A*, *BC* and *CA* represent two forces, which, acting along *bc* and *ca* respectively, would be in equilibrium with *P*.

Draw *CD*, *AD* parallel to *cd*, *ad* respectively, to meet in *D*. Then *CD*, *DA* represent two forces, which, acting along *cd*, *da* respectively, would be equivalent to the force represented by *CA* acting along *ca*.

Hence, measuring *BC*, *CD*, *DA*, we have the three forces required.

If the three lines *bc*, *cd*, *da* are parallel to one another but not to the direction of *P*, the problem is impossible of solution. In this case *oc* is in the same straight line with *bc*, and the line *BC* does not meet the line *OC*.

If two of the three given lines are each parallel to the line of action of P, the force along the third line vanishes. For instance, if ab, bc, cd are parallel, C is a point in AB, and D coincides with A.

If all four lines are parallel, the solution is indeterminate. This case will be considered in the next chapter.

107. *To resolve a given force into three components along given lines of action situated in one plane.*

If we suppose the given force to be reversed in direction, it will form a system in equilibrium with the three components required. Hence this problem reduces to the preceding.

108. Ex. 1. *A triangular lamina ABC, of no appreci-*

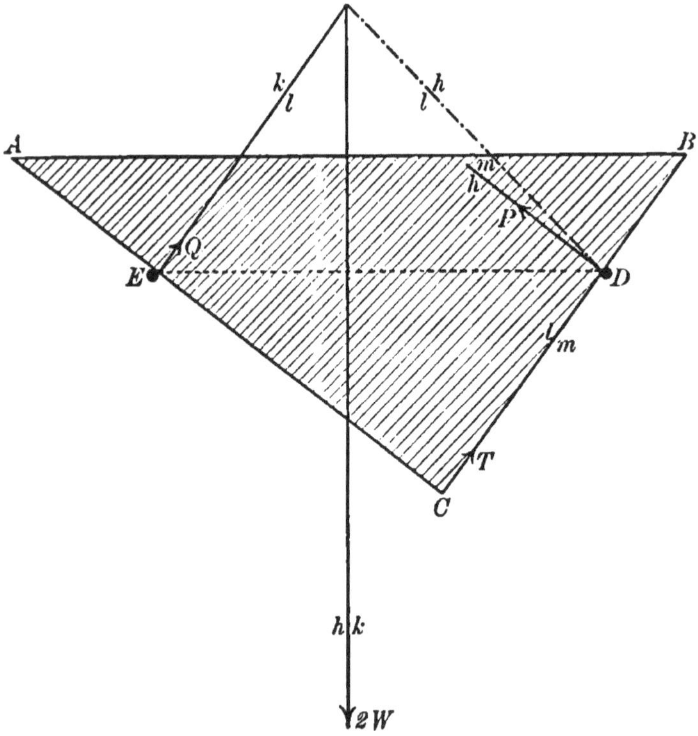

FIG. 117.

able weight, whose sides BC, CA, AB are respectively
18, 24, and 30 inches in length, is placed in a vertical

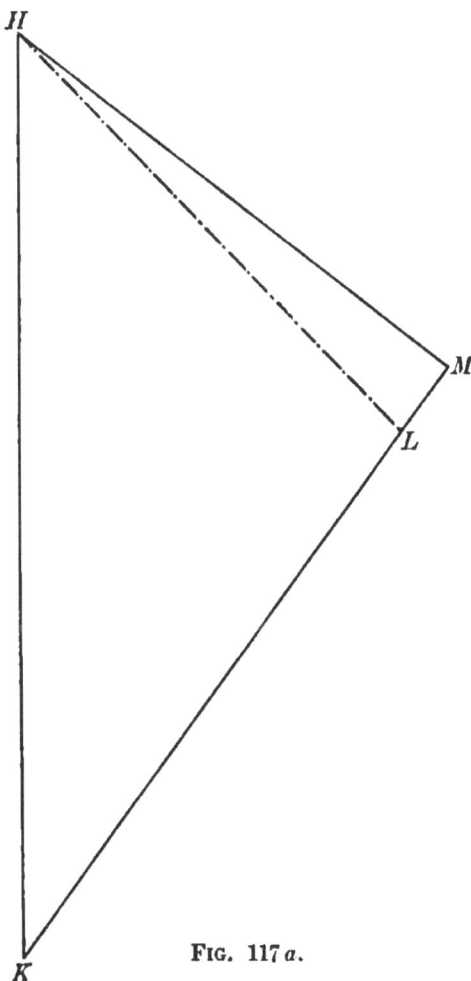

FIG. 117 a.

plane with BC, CA resting upon two fixed smooth pegs
D and E, situated 20 inches apart in the same horizontal
line. If masses, each of weight W, are suspended from

*A and B, and the triangle is kept with AB horizontal
by means of a fine light string connecting C with the
peg D, find the tension of the string and the pressures
on the pegs.*

The space diagram is easily constructed to scale.

Consider the forces acting on the lamina. The two
equal forces W, acting vertically downwards at A and
B, can be replaced by a single force $2W$, acting vertically downwards through the middle point of AB, along
a line which we draw and mark hk.

Choosing any suitable length to represent W, we draw
HK vertically downwards of length to represent $2W$.

Let P and Q be the pressures of the pegs D and E
respectively upon the lamina; these are perpendicular
to CB, CA respectively. Let T be the tension of the
string CD.

The four forces $2W$, Q, P, T, acting on the lamina,
are in equilibrium. We see that the lines of action of
the first two of these forces meet in an accessible point,
and the other two act through D. Hence, marking the
lines of action of Q, T, P with the letters kl, lm, mh
respectively, we draw the straight line hl, connecting D
with the point of intersection of hk and kl.

Draw straight lines HL, KL parallel to hl, kl respectively, thus obtaining the point L. Draw straight lines
LM, HM parallel to lm, hm respectively, and we have
the point M.

Then KL, LM, MH represent Q, T, P respectively.
On measuring these lines we find that

$$Q = (1\cdot43)W,$$
$$T = (\cdot18)W,$$
$$P = (1\cdot20)W.$$

109. Ex. 2. *Four forces in equilibrium act in the lines AB, BC, CD, DA. From any point A′ in BD, or BD produced either way, a straight line is drawn parallel to AB to meet AC in B′; from B′ a straight line is drawn parallel to BC to meet BD in C′; from C′ a straight line is drawn parallel to CD to meet AC in D′; and the straight line joining D′, A′ is drawn. It is required to prove that D′A′ is parallel to DA, and that the forces in the lines AB, BC, CD, DA are proportional to A′B′, B′C′, C′D′, D′A′ respectively.*

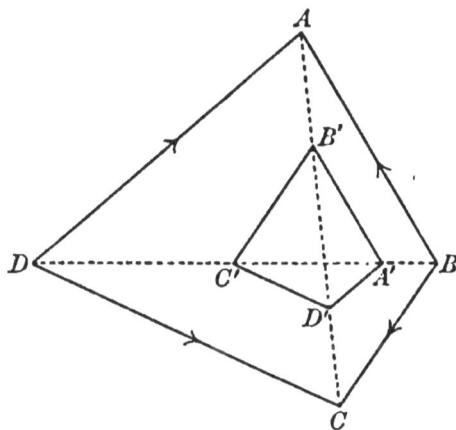

FIG. 118.

Let the scale of representation of force be so chosen that $A'B'$ may represent the force which acts in the line AB. If the force in AB acts in the opposite direction to $A'B'$, we may suppose all four forces to be reversed; they will still form a system in equilibrium, provided their magnitudes are unaltered.

Since the resultant of the forces in AB, BC is a force in the line BD, it follows that $B'C'$ must represent the force in BC. Also, since the resultant of the forces in BC, CD is a force in the line CA, we see that $C'D'$

must represent the force in CD.   Hence $D'A'$, which closes the polygon $A'B'C'D'$, must represent the force in DA.

Therefore $D'A'$ is parallel to DA, and the forces in the lines AB, BC, CD, DA are proportional to $A'B'$, $B'C'$, $C'D'$, $D'A'$ respectively.

Also, the directions of the forces in the lines AB, BC, CD, DA are determined by the direction arrows going ·one way round the quadrilateral $A'B'C'D'A'$. In the figure given above, the directions of the forces are BA, BC, DC, DA, or, reversing these directions, AB, CB, CD, AD.

The student should make himself familiar with the generality of this proposition.   In the first of the two figures given below the directions of the forces are AB, BC, DC, AD, or these directions reversed; in the second figure they are one way round.

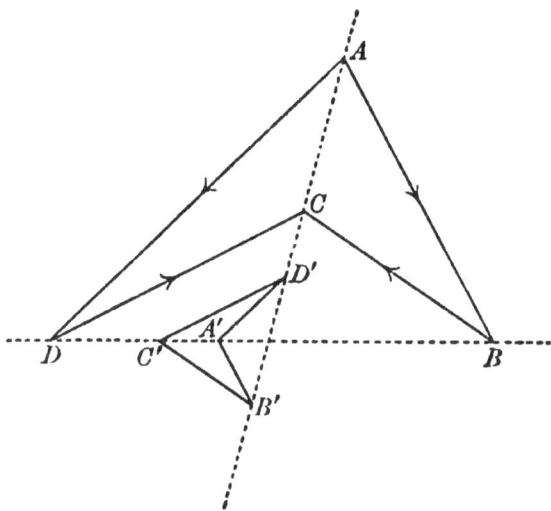

FIG. 119.

As the point A' may be taken anywhere in BD, or BD produced either way, the student should try the effect of making A coincide with B or D. He should also notice that the figure $ABCD$ has the same relation to $A'B'C'D'$ as $A'B'C'D'$ has to $ABCD$. Hence, if four

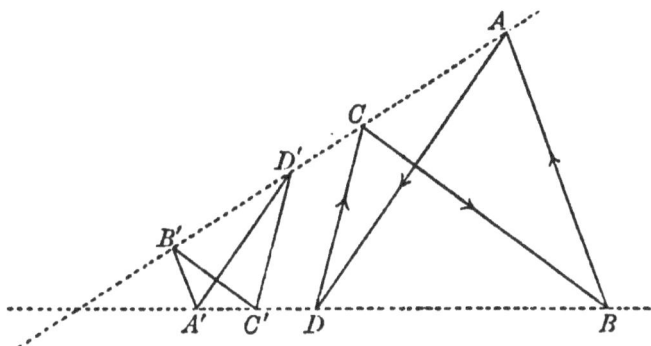

FIG. 120.

forces in equilibrium act in the lines $A'B'$, $B'C'$, $C'D'$, $D'A'$, the figure $ABCD$ may be taken to be the force diagram for the system.

EXAMPLES XI.

1. A uniform thin rod ACB, of length 3 feet 9 inches, and weighing 25 pounds, rests with the lower end A upon a smooth horizontal plane, and against the smooth edge of a step, 1 foot 6 inches high, at C. It is kept from slipping by a fine light string, 2 feet long, connecting A with a point at the foot of the step vertically below C. Find all the external forces acting on the rod.

2. A straight uniform rod AB, of mass 30 pounds, rests against a smooth horizontal plane at A and against a smooth fixed rail at C, where $BC=\frac{1}{4}BA$. It is prevented from slipping by a fine light string AD, connecting A with a point D, situated in the horizontal plane vertically below C. If the angle BAD is 30°, find the tension of the string, and the pressures at A and C.

If the plane be rough, and there is no string, what must be the coefficient of friction between the rod and the plane in order that the rod may be just on the point of slipping?

3. A square board $ABCD$ is placed upon a smooth horizontal table, and a given force P acts from E, the middle point of AB, towards F, the middle point of CD. Determine the magnitudes of three forces X, Y, Z, which, acting along BC, CA, AB respectively, will make equilibrium with P.

4. The lower extremity E of a uniform beam ED rests on the ground at the foot of a vertical wall EF, its upper extremity being attached by a fine light cord DF to a point F, situated vertically above E. The mass of the beam being 200 pounds, find the tension of the cord, and the pressures of the beam against the wall and ground at E, supposing that $EF=2FD=ED$.

5. A uniform beam AB, of mass 100 pounds, is supported by two fine light strings AC, BD, the latter being vertical, and the angles DBA and BAC being 100° and 130° respectively. The beam is maintained in this position by a horizontal force of P pounds' weight, applied at B. Find the value of P.

6. A uniform square lamina $ABCD$, weighing 10 pounds, is constrained at A and B to remain in contact with a smooth fixed straight vertical rod, A being uppermost. The point C rests in contact with a smooth fixed plane, inclined at an angle of 60° to the horizon. Find the actions at A, B, C.

7. $ABCD$ is a fine straight rod of no appreciable weight, the portions AB, BC, CD being of lengths 9, 6, 5 inches respectively. A mass of 10 pounds is suspended from A, and a mass of unknown weight is suspended from D. The rod is supported in a horizontal position by means of two forces applied at B and C in directions BE, CF respectively, the angles ABE and DCF being each 80°. Determine the weight of the mass suspended from D, and the magnitudes of the forces applied at B and C.

8. A lamina $ABCD$, having DC parallel to AB, and such that $AD=DC=CB=\frac{1}{2}AB$, is placed upon a smooth horizontal table, and a given force P acts from A towards B. Find the magnitudes of three forces X, Y, Z, which, acting along AD, CD, CB respectively, will make equilibrium with P.

9. $ABCDEF$ is a regular hexagon. Find what forces must act along AC, AF, DE, to produce equilibrium with a force of 40 pounds' weight acting along EC.

10. A straight bar ACB, of length 5 feet, and of no appreciable weight, is supported in a horizontal position with a load of 50 pounds applied at C, 2 feet from A, by means of two strings AH, BK, such that the angles BAH and ABK are 120° and 150° respectively. Find what force must be applied at D, the middle point of AC, in direction DL, in order that the bar may remain horizontal ; DL being downwards, and the angle ADL being 60°. Also, determine the tensions of the strings.

11. A uniform rectangular lamina $ABCD$, weighing 100 pounds, rests in a vertical position upon a smooth horizontal plane at A, and against a smooth vertical wall at D. It is supported by a fine string FC attached to a point in the wall. If, in the position of equilibrium, the triangle FCD is equilateral, find the tension of the string, and the pressures at D and A, given that $DC = 2DA$.

12. A uniform beam AB, weighing 100 pounds, is supported by strings AC, BD, the latter being vertical. It is maintained in this position by a horizontal force P applied at B. Find the value of P in pounds' weight, the angles CAB, ABD being each 105°.

13. A ladder AB, of length 30 feet, inclined at an angle of 60° to the horizon, rests against a smooth wall BC, inclined at 75° to the horizon, and upon a smooth horizontal plane AC. The end A is kept from slipping by a fine light string, connecting it with the point C. If the centre of gravity of the ladder, which weighs 40 pounds, is 12 feet from A, find the tension of the string, and the pressures at A and B.

14. $AHKB$ is a straight rod, of no appreciable weight, and of length 10 feet, the points H and K being 1 foot from A and 6 feet from B respectively. At H and K are suspended two masses P and Q respectively, and the rod rests in a horizontal position, being supported by two fine light strings AC, BD, such that the angles BAC, ABD are each 150°. If P is 10 pounds, find Q, and the tension of each string.

15. D is the orthocentre of the triangle ABC, whose sides BC, CA, AB are of lengths 14, 13, 15 inches respectively. Find what forces must act along the lines CB, DC, DA to be in equilibrium with a force of 25 pounds' weight acting along AB.

16. Four forces in equilibrium act in the lines AB, BC, CD, DA. $A\delta$ is drawn parallel to BC to meet BD in δ, and $B\gamma$ is drawn parallel to AD to meet AC in γ. Prove that $\gamma\delta$ is parallel to CD; also, that the forces in the lines AB, BC, CD, DA are proportional to AB, δA, $\gamma\delta$, $B\gamma$ respectively.

17. Three forces P, Q, R, acting along the lines BC, CA, AB respectively, are in equilibrium with a force X acting along a line drawn parallel to BC through a point K in BA produced. A straight line through A, parallel to KC meets BC in H. Prove that $P : Q : R : X = HC : CA : AB : BH$.

18. Three forces P, Q, R, acting along the lines BC, CA, AB respectively, are in equilibrium with a force X acting through K, a point in BA produced. Prove that, whatever be the direction of the force X, the ratio $P : Q$ is constant.

19. A given force, represented by AB, acts along a given line ab, and is in equilibrium with three unknown forces, acting along three given lines bc, cd, da. The lines cd, da intersect at the point δ, and a straight line through δ intersects ab, bc in a and γ respectively. Prove the following method for determining the magnitudes of the three unknown forces:

Divide AB in C' in the same ratio that a divides $\gamma\delta$; let straight lines through C' and B, parallel to $\gamma\delta$ and bc respectively, meet in C; also let straight lines through C and A, parallel to cd and da respectively, meet in D. Then BC, CD, DA represent the magnitudes of the forces which act along the lines bc, cd, da respectively.

Apply the method to the solution of Question 5.

20. A uniform beam AB is supported by two fine light strings AC, BD, the latter being vertical, and the angles DBA, BAC equal to one another. The beam is maintained in this position by a horizontal force applied at B. Show that the tension of the string AC is equal to half the weight of the beam.

21. In the figure of Art. 103, prove that BD is parallel to the line joining the intersection of bc and cd with the intersection of ab and da.

22. In the figure of Art. 103, if the parallelogram $BADA'$ be completed, prove that CA' is parallel to the line joining the intersection of ab and cd with the intersection of bc and da.

23. A force along AB is in equilibrium with three forces in the lines BC, CD, DA ; prove each of the following :

(i.) If AB, CD are parallel, and in the same direction, the forces are in directions AB, BC, CD, DA, and are proportional to CD, BC, AB, DA respectively.

(ii.) If AB, CD are parallel, and in opposite directions, the forces are in directions AB, CB, CD, AD, and are proportional to CD, BC, AB, DA respectively.

(iii.) If AC, BD, are parallel, and either in the same or opposite directions, the forces act one way round, and are proportional to the lines along which they respectively act.

(iv.) If $ABDC$ is a parallelogram, or if $ABCD$ is a parallelogram, the forces are proportional to the sides along which they respectively act.

(v.) If A, B, C, D are points taken one way round on the circumference of a circle, the forces are in directions AB, CB, CD, AD, and are proportional to CD, DA, AB, BC respectively.

(vi.) If A, B, D, C are points taken one way round on the circumference of a circle, the forces are in directions AB, BC, CD, DA, and are proportional to CD, DA, AB, BC respectively.

(vii.) If one of the points A, B, C, D is the orthocentre of the triangle formed by the joins of the other three, the forces in the lines AB, BC, CD, DA are proportional to CD, DA, AB, BC respectively.

(viii.) If D is the intersection of the medians of the triangle ABC, the forces are in directions AB, CB, DC, DA, and are proportional to AB, BC, $3CD$, $3DA$ respectively.

24. Prove the following geometrical properties of the figures of Art. 109 :

(i.) If D is the centre of the circle which circumscribes the triangle ABC, prove that B' is the centre of a circle which touches the sides of the triangle $C'D'A'$.

(ii.) If D is the centre of a circle which touches the sides of the triangle ABC, prove that B' is the centre of the circle which circumscribes the triangle $C'D'A'$.

(iii.) If one of the points A, B, C, D is the orthocentre of the triangle formed by the joins of the other three, prove that each of the points A', B', C', D' is the orthocentre of the triangle formed by the joins of the other three.

Prove also that in this case the figures $ABCD$, $C'D'A'B'$ are similar.

(iv.) If the points A, B, C, D are concyclic, so are the points A', B', C', D'.

Prove that in this case also the figures $ABCD$, $C'D'A'B'$ are similar.

(v.) If the figure $A'B'C'D'$ is similar to the figure $CDAB$, prove that *either* the four points A, B, C, D are concyclic, *or* each is the orthocentre of the triangle formed by the joins of the other three.

(vi.) If the straight lines AC, BD intersect at O, prove that
$$OA \cdot OA' = OB \cdot OB' = OC \cdot OC' = OD \cdot OD'.$$
Hence show that if distances OA'', OB'', OC''', OD'' are taken along OA, OB, OC, OD respectively, such that
$$OA \cdot OA'' = OB \cdot OB'' = OC \cdot OC''' = OD \cdot OD'',$$
the forces in AB, BC, CD, DA are proportional to $A''B''$, $B''C''$, $C''D''$, $D''A''$ respectively.

(vii.) If $OA \cdot OB = OC \cdot OD$, prove that the forces in AB, CD are proportional to AB, CD respectively.

(viii.) If the parallelogram $B'A'D'a$ be completed, prove that $C'a$ is parallel to the line joining the intersection of BC and DA to the intersection of CD and AB.

25. Four forces in equilibrium act in the lines AB, BC, CD, DA. Points A', B', C', D' are the centres of the circles which circumscribe the triangles BCD, CDA, DAB, ABC respectively. Prove that the forces in the lines AB, BC, CD, DA are proportional to $C'D'$, $D'A'$, $A'B'$, $B'C'$ respectively.

26. Four forces in equilibrium act along straight lines, which bisect AB, BC, CD, DA at right angles; prove that their directions must be such that, in going along $ABCDA$, the forces are all to the right or all to the left, and that the magnitudes of the forces are proportional to the lines to which they are respectively perpendicular.

27. Two known forces, represented by AB, BC, act along known lines ab, bc respectively, and are in equilibrium with two unknown forces, one of which acts along a known line cd, and the other through a known point H. Prove the following method for determining the unknown forces :

Through the intersection of ab, bc draw ac parallel to AC, and from H draw da to the point of intersection of ac, cd. Through A and C draw straight lines parallel to ad, cd respectively, to meet in D. Then CD, DA represent the unknown forces, and cd, da are their lines of action respectively.

28. In the preceding example, prove also the following method for determining the unknown forces :

Through H draw ax to the point of intersection of ab and bc, and let the straight line through A, drawn parallel to ax, meet BC in X. Through H draw xd to the point of intersection of bc and cd, and let straight lines through X and C, drawn parallel to xd, cd respectively, meet in D. Join DA, and draw through H a straight line da parallel to DA. Then CD, DA represent the unknown forces, and cd, da their lines of action respectively.

CHAPTER XII.

EQUILIBRIUM OF PARALLEL FORCES IN ONE PLANE.

110. We have seen that, in order to *insure* equilibrium, it is necessary and sufficient that a *force polygon* and a *funicular polygon* corresponding to it should both close.

The general method of work in solving problems is as follows: A rigid body is in equilibrium under the

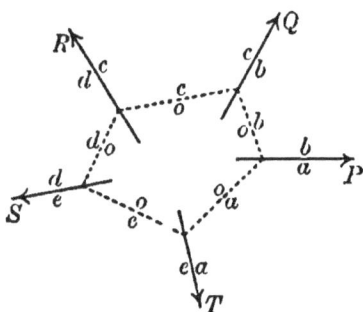 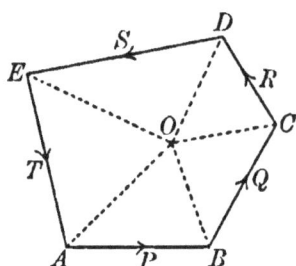

FIG. 121. FIG. 121 *a.*

influence of a number of forces, some partly or wholly known, and others partly or wholly unknown. From the data we construct as much of the force polygon and funicular polygon as we can, planning out first those forces which are wholly known, and then we endeavour to complete both polygons, making them both close.

We append the space diagram and force diagram for a system of five forces P, Q, R, S, T in equilibrium. The student cannot make himself too familiar with the manner in which the two figures correspond.

111. In this chapter we confine ourselves to the case in which the forces are parallel. The force polygon reduces to a straight line, or rather, a double straight line. Let a system of parallel forces whose measures are P, Q, R, X, Y be in equilibrium. Let $ABCDEA$ be the force polygon for the system, so that AB, BC, CD, DE, EA represent the forces P, Q, R, X, Y respectively, and let the lines of action of these forces be marked ab, bc, cd, de, ea respectively.

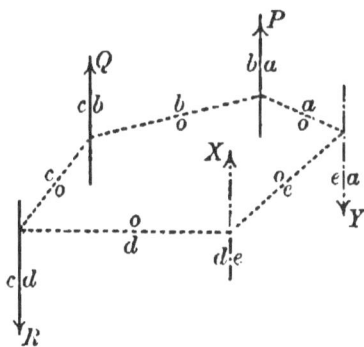
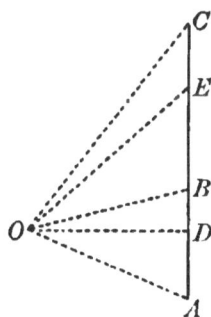

FIG. 122.   FIG. 122 a.

Suppose that the forces P, Q, R are known completely, but that the forces X and Y are at present partly or wholly unknown. We are able to construct the force polygon to this extent: we can with any suitable scale construct $ABCD$, but the point E is at present unknown.

Take any pole O and draw OA, OB, OC, OD. We can then construct the funicular polygon to this extent: we can draw the strings oa, ob, oc, od, the point of intersection of oa and ob being chosen anywhere in ab, but the string oe cannot at present be drawn.

Thus we see that there are three points to be determined,—$viz.$, the point E and the extremities of the string oe,—and these points are subject to this condition, that OE shall be parallel to oe. The position of E determines the magnitudes and directions of X and Y.

The work now falls under the following cases:

I. Let the magnitude and direction of one of the two forces X and Y be known, and the line of action of one of them also known.

The knowledge of the magnitude and direction of one gives the point E immediately, and determines the magnitude and direction of the other also. The knowledge of the line of action of one (say X) gives one extremity of the string oe, for we can find the point of intersection of od and de. We can then draw oe parallel to OE, and the point of intersection of oa and oe is a point in the line of action of the remaining force (Y).

II. Let the lines of action of the two forces X and Y be known. Then we have at once both extremities of the string oe. We then draw OE parallel to oe, and thus determine the position of E, and with it the magnitudes and directions of X and Y.

112. As an example of the above, let it be required to find three forces' acting along the given lines bc, cd, da, which are in equilibrium with a given force whose measure is P acting along the given line ab; the straight lines ab, bc, cd, da being all parallel.

Take AB to represent the given force, and in AB take *any* point C. We can then find, by the method given above, two forces acting along cd, da, which are in equilibrium with the given force and a force represented by BC acting along bc.

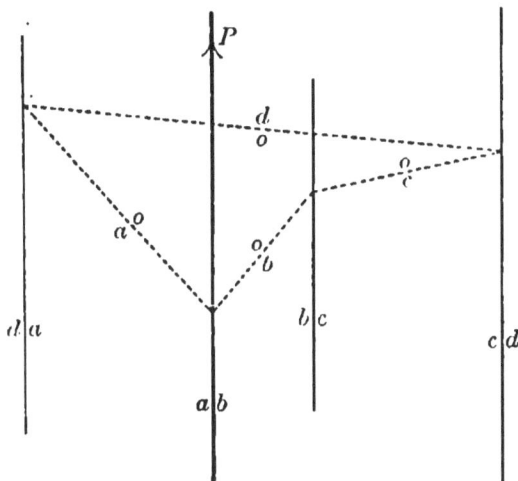
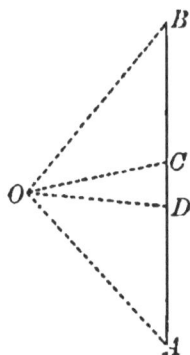

FIG. 123. FIG. 123 a.

Thus the problem is indeterminate, as we may take C *anywhere* in AB.

So, also, the problem of resolving a given force into three parallel forces, along known lines parallel to its own direction, is indeterminate.

113. Ex. 1. *A rigid beam, acted upon by a given system of vertical forces, rests in a given horizontal position, being supported by two smooth pegs, situated at given points. It is required to determine the pressures between the beam and the pegs.*

Let AB, BC, CD, DE be taken to represent the given forces, and let their lines of action be marked ab, bc, cd, de respectively.

The pressures of the pegs upon the beam will be both vertical. Let the verticals through the pegs be called ef, fa.

Take any pole O, and draw OA, OB, OC, OD, OE. Starting from any point in ab, we can draw the strings oa, ob, oc, od, oe of the funicular polygon. Also, joining

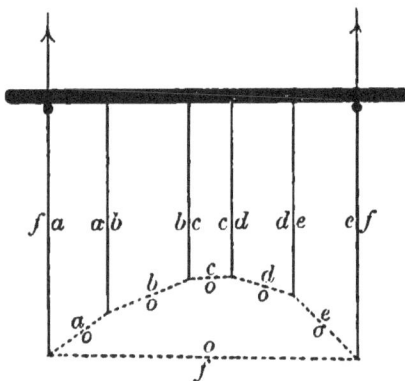 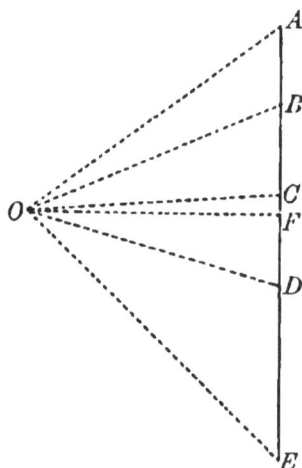

FIG. 124.   FIG. 124 a.

the point of intersection of oa and af with the point of intersection of oe and ef, we have the string of completing the funicular polygon.

Draw OF parallel to of, to meet AE in F. Then EF, FA represent the pressures of the pegs upon the beam acting along ef, fa respectively.

114. Ex. 2. *A fine rod AB, whose centre of gravity is in a given position G, and whose weight is of given magnitude w, rests in a horizontal position ·upon two smooth pegs C and D, situated in given positions. The pegs C and D cannot sustain pressures greater than P_0 and Q_0 respectively. It is required to find the*

*portion of the rod, at any point of which a given
load W may be applied.*

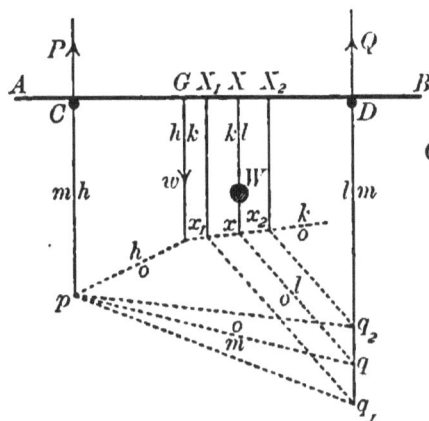
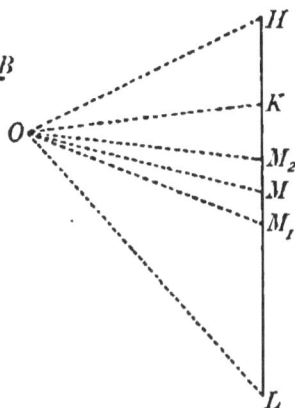

FIG. 125.   FIG. 125 a.

Suppose the rod is in equilibrium when the load W
is applied at X. Let P and Q be the pressures of the
pegs C and D respectively upon the rod.

Draw HK, KL to represent w and W respectively,
and let the verticals through G, X, D, C be marked
hk, kl, lm, mh respectively. Take any pole O.

From any point p in mh draw the string oh, and
from the point of intersection of oh and hk draw the
string ok, meeting kl in x. From x draw the string
ol, meeting lm in q. Then, joining pq, we have the
string om, completing the funicular polygon. Draw
OM parallel to om, to meet HL in M. Then LM, MH
represent Q and P respectively.

For different positions X_1 and X_2 of X, we have
different positions x_1 and x_2 of x, which give different
positions q_1 and q_2 of q; and these lead to different
positions M_1 and M_2 of M.

Measure HM_1 vertically downwards to represent P_0, and LM_2 vertically upwards to represent Q_0. Then, unless HM_1 and LM_2 overlap, it will be impossible to support the load. If they *do* overlap, the point M must lie between M_1 and M_2.

By drawing pq_1, pq_2 parallel to OM_1, OM_2 respectively, we easily get the points q_1 and q_2, and from these the points x_1 and x_2, and then the points X_1 and X_2. Hence X must be some point between X_1 and X_2. Thus X_1X_2 is the required portion of the rod.

115. Ex. 3. *A fine heavy rod ACDB rests in a horizontal position upon two smooth pegs C and D, situated in given positions. The greatest load ·which can be applied at A without disturbing the equilibrium is of given weight P, and the greatest load which can be applied at B without disturbing the equilibrium is of given weight Q. ♦It is required to determine the weight of the rod, and the position of its centre of gravity.*

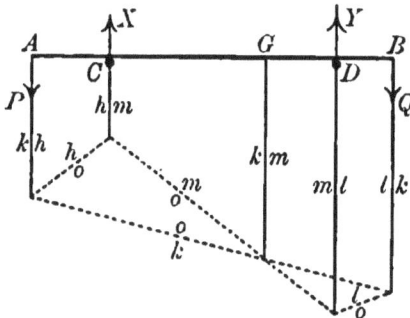

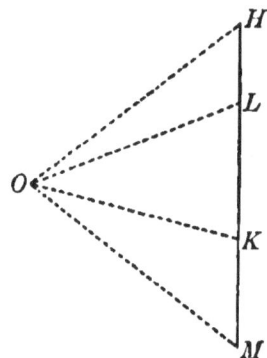

FIG. 126.   FIG, 126 a.

When the load P is applied at A, the rod is on the point of turning about C; thus there is no pressure between the rod and the peg D, and the pressure of the peg C upon the rod is an unknown force X vertically upwards.

When the load Q is applied at B, there is no pressure at C, and the pressure of the peg D upon the rod is an unknown force Y vertically upwards.

The weight of the rod balances the two forces P and X; it also balances the two forces Q and Y. Hence, if we reverse the two forces Q and Y, the four forces P, X, Q reversed and Y reversed form a system in equilibrium.

Draw HK, KL to represent the force P and the reversed force Q respectively, and mark the verticals through A and B with the letters hk, kl respectively; also, let the verticals through D and C be called lm, mh respectively.

Take any pole O, and draw OH, OK, OL. Starting from any point on hk, draw the strings oh, ok, ol, and complete the funicular polygon by drawing the string om. Draw OM parallel to om, to meet HK produced in M. Then MH represents X.

Now the forces represented by MH, HK, acting along the lines mh, hk respectively, are in equilibrium with the weight of the rod. Therefore the weight of the rod is represented by KM, and acts along a vertical line through the intersection of om and ok. Hence, drawing a vertical line through the intersection of the strings om, ok to meet the rod in G, we see that the point G so determined is the centre of gravity of the rod, whose weight is represented by KM.

EXAMPLES XII.

1. A bookshelf, supported at its extremities, is just filled by two sets of books, the books of each set being placed together. One set consists of 14 volumes, each $1\frac{1}{2}$ inches thick, and each weighing $2\frac{1}{2}$ pounds; the other consists of 12 volumes, each $1\frac{1}{4}$ inches thick, and each weighing 2 pounds. Find the pressures on the supports, the mass of the shelf being 8 pounds.

2. A bent lever, of weight W, consists of two uniform, heavy, straight rods, whose lengths are as 3 to 4; find the weight of the mass, which must be attached to the end of the shorter rod, in order that—the fulcrum being at the junction of the two rods, which are of the same material and thickness—they may make equal angles with the horizon.

3. A uniform straight rod, of length 18 inches, and weighing 9 pounds, is suspended from its extremities by two vertical strings, neither of which can support a tension greater than 50 pounds' weight. Find the greatest load which may be applied to the rod at a point 5 inches from one end.

4. A heavy uniform beam, of length 20 feet, and weighing 50 pounds, is suspended horizontally by two vertical strings attached to its extremities, each of which can sustain a tension of 40 pounds' weight. How far from the centre of the beam must a mass of 20 pounds be placed, so that one of the strings may just break?

5. A uniform rod AB, of length 1 foot, and mass 10 pounds, is suspended at A and B, in a horizontal position, by two vertical strings, each of which can support a tension of 26 pounds' weight; how far from the centre of the rod must a mass of 28 pounds be placed, so that one of the strings may just break?

6. A heavy straight rod $ACDB$, of length 12 inches, rests upon two smooth pegs C and D, distant 3 inches and 2 inches respectively from A and B. The greatest loads which can be applied in turn at A and B, without disturbing the equilibrium, are 8 and 9 pounds respectively. Find the weight of the rod, and the position of its centre of gravity.

7. A heavy straight rod AB, of length 15 inches, balances about a point 2 inches from A, when a mass of 6 pounds is suspended from A. It balances about a point 2 inches from B, when a mass of 5 pounds is suspended from B. Find the weight of the rod and the position of its centre of gravity ?

8. A heavy uniform bar $ACDB$ rests in a horizontal position upon two fixed supports C and D, whose distance apart is 6 inches, and equal to the length of the projecting part AC of the bar. If an upward force of 2 pounds' weight, applied at A, just lifts the bar off the support C, and a downward force of 8 pounds' weight at A just lifts it off D, find the length and weight of the bar.

9. A rod ABC, 16 inches long, rests in a horizontal position upon two supports at A and B, one foot apart, and it is found that the least upward and downward forces applied at C, which would move the rod, are 4 ounces' weight and 5 ounces' weight respectively. Find the weight of the rod, and the position of its centre of gravity.

CHAPTER XIII.

EQUILIBRIUM OF COPLANAR FORCES.

116. We do not propose to discuss here in detail the different cases which may arise in solving problems on the equilibrium of a system of forces acting upon a rigid body in one plane. It will suffice to discuss the following two cases of frequent occurrence.

117. I. *A number of forces in one plane are in equilibrium. All are known completely with the exception of two. Of these, the line of action of one is known, and the point of application of the other. It is required to determine the unknown forces completely.*

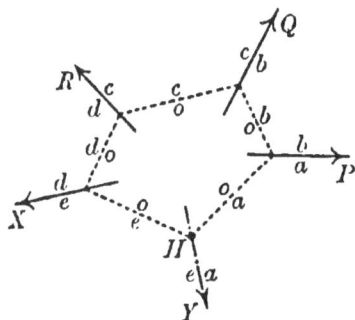 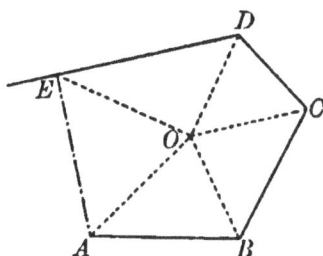

FIG. 127.   FIG. 127 a.

Let P, Q, R, X, Y be the measures of five forces in equilibrium, P, Q, R being all known and X and Y at present unknown. Let the lines of action of P, Q, R, X be also given and represented by those indicated in the figure on the left, and let H be a given point in the unknown line of action of the force Y.

Draw lines AB, BC, CD to represent in magnitude and direction the given forces P, Q, R respectively. Draw also DE in the direction of X. Mark the lines of action of the forces P, Q, R, X by the letters ab, bc, cd, de respectively.

To complete the force polygon we have only to find the remaining vertex, which is somewhere in DE, and will be denoted by E. The line EA will represent Y, and its line of action will be denoted by ea, but at present the position of E and the direction of ea are unknown.

· Take any pole O and draw the rays OA, OB, OC, OD. The remaining ray OE cannot at present be drawn.

As H is the only point known in the line of action of Y, we will commence to construct our funicular polygon at H. Draw through H a straight line ao parallel to AO. From the point of intersection of oa and ab draw ob parallel to OB. From the intersection of ob and bc draw oc parallel to OC. From the intersection of oc and cd draw od parallel to OD. To complete the funicular polygon we have merely to join H to the point of intersection of od and de. The line so drawn we call oe, and it must be parallel to the remaining ray of the force polygon.

Hence draw OE parallel to oe to meet DE in E, and join E, A. Then DE gives us X, and EA gives the magnitude and direction of Y.

118. II. *A number of forces in one plane are in equilibrium. All are known completely with the exception of three whose lines of action are known. It is required to determine the magnitudes of the unknown forces.*

We can deduce this from the preceding case. Let P, Q, R, X_1, X_2, X_3 be the measures of six forces acting along known lines, and keeping a rigid body in equilibrium, the three P, Q, R being known, and the three X_1, X_2, X_3 at present unknown.

Find H, the point of intersection of the lines of action of X_2 and X_3. Then the two forces X_2 and X_3 are equivalent to an unknown force whose measure is Y (say) acting through H in an unknown direction.

Hence we proceed exactly as in the preceding case. Having determined in this way the position of E, we know that EA represents Y. Then, denoting the lines of action of X_2 and X_3 by ef, fa respectively, we draw through E and A straight lines parallel to ef, af respectively to meet in F.

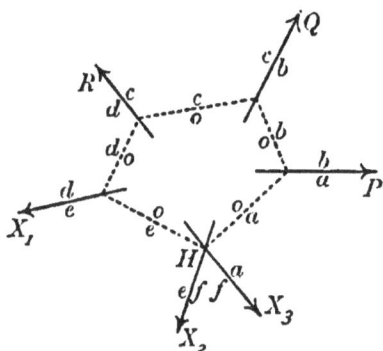

FIG. 128.

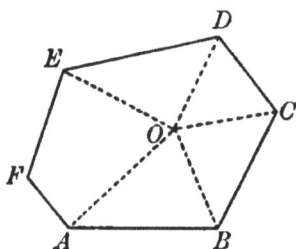

FIG. 128 *a*.

The three unknowns X_1, X_2, X_3 are found by measuring DE, EF, FA respectively.

If no two of the lines of action of X_1, X_2, X_3 meet in an accessible point, we may proceed as follows:

Before choosing the position of the pole O, draw from a point in ab a straight line oa towards the inaccessible point of intersection of the lines of action of X_2 and X_3. This we can do by the method described in Art. 104.

Through A draw AO parallel to ao, in AO choosing any point O, and draw the rays OB, OC, OD. Then we may proceed as before, completing the funicular polygon by drawing a straight line oe from the point of intersection of od and de towards the inaccessible point of intersection of the lines of action of X_2 and X_3.

119. *Ex. A uniform ladder, weighing 85 pounds, and of length 20 feet, rests with one end against a smooth vertical wall and the other end upon the smooth horizontal ground. A man weighing 150 pounds stands on the ladder at a point three-quarters of the way up, and it is kept from slipping by a fine horizontal string of length 9 feet, attached to the ladder at a point a quarter of the way up, and to the wall at a point vertically below the top of the ladder. Find the tension of the string and the reactions of the ground and the wall.*

We can construct the space diagram to scale from the data. Let AB represent the ladder, D its middle point, C and E the middle points of AD and DB respectively, CF the horizontal string attached to the ladder at C and to the wall BG and F.

Consider the external forces acting on the ladder and man as one system. They are:—85 pounds' weight acting vertically downwards through D, 150 pounds' weight acting vertically downwards through E, the tension (T pounds' weight) of the string along CF, the reaction of the ground (R pounds' weight) vertically upwards through A, and the reaction of the wall (S pounds' weight) acting horizontally through B. Thus we have an example of Case II. above.

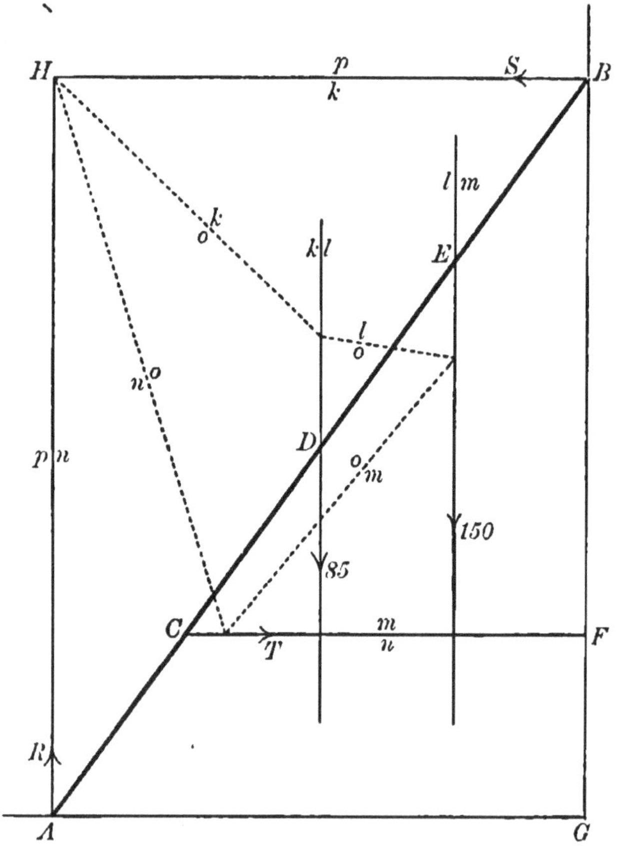

FIG. 129.

Let the lines of action of R and S intersect at H. Then R and S are equivalent to an unknown force acting through H.

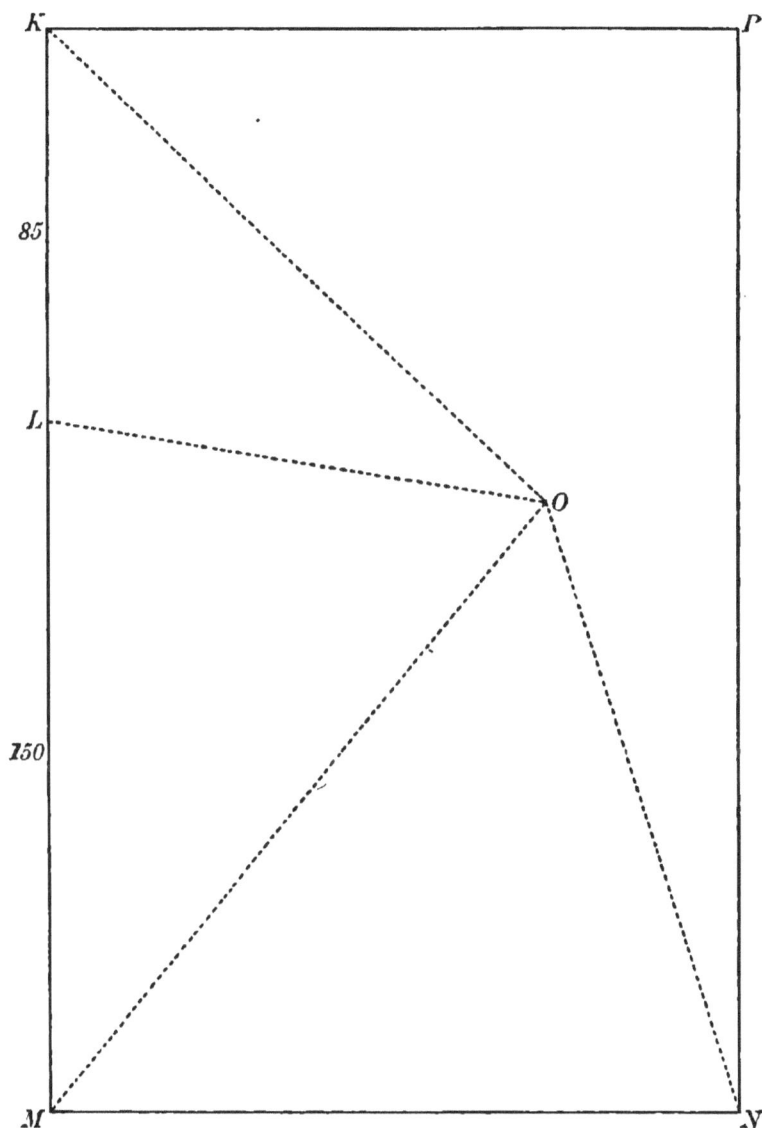

FIG. 129 a.

With any suitable scale draw KLM vertically downwards, making KL, LM of lengths 85 and 150 units respectively, and mark the verticals through D and E

with the letters kl, lm respectively; also mark the lines of action of T, R, S with the letters mn, np, pk respectively. Take any pole O and draw the rays OK, OL, OM. Through H draw ok parallel to OK; through the point of intersection of ok and kl draw ol parallel to OL; through the point of intersection of ol and lm draw om parallel to OM; through H draw the line on to the point of intersection of om and mn, thus completing the funicular polygon. Draw ON parallel to on to meet the horizontal through M in N, and let the vertical through N meet the horizontal through K in P. Then $KLMNPK$ (this way round) is the force polygon.

On measuring the lines MN, NP, PK we find that $T = 155$, $R = 235$, $S = 155$. Thus

$$\left\{ \begin{array}{l} \text{Tension of string} = 155 \text{ pounds' weight.} \\ \text{Reaction of ground} = 235 \text{ pounds' weight.} \\ \text{Reaction of wall} = 155 \text{ pounds' weight.} \end{array} \right.$$

Otherwise.—The force polygon $KLMNPK$ will evidently be a rectangle. This shows at once that $R = 235$ and $S = T$.

Now the weight of the ladder is equivalent to two forces each of $42\frac{1}{2}$ pounds' weight acting vertically downwards through A and B. The weight of the man is equivalent to a load of $37\frac{1}{2}$ pounds at A and $112\frac{1}{2}$ pounds at B. Also the tension of the string is equivalent to $\frac{3}{4}T$ along AG and $\frac{1}{4}T$ along Hb.

Considering only those forces which may be taken as acting at A, we see that we have

$R - 42\frac{1}{2} - 37\frac{1}{2}$, *i.e.* 155 pounds' weight, vertically upwards, and $\frac{3}{4}T$ along AG.

These two forces balance forces at B. Therefore their resultant acts in the line AB.

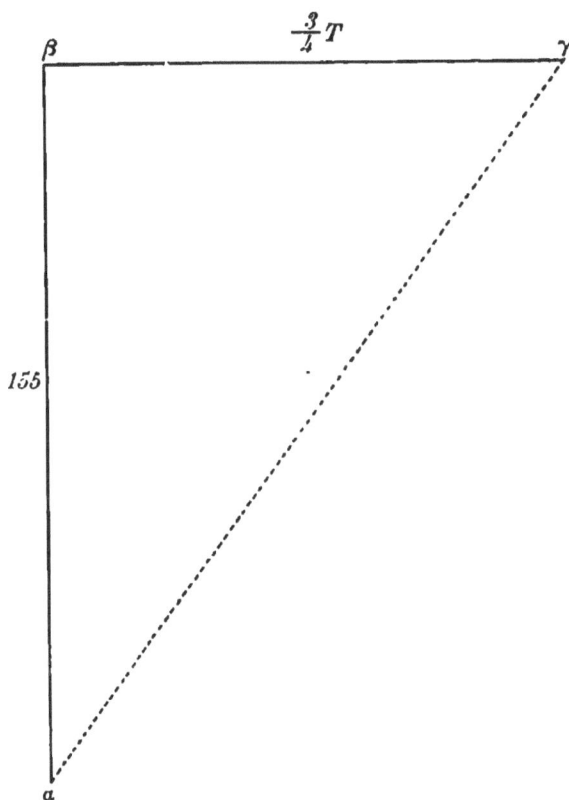

FIG. 129 b.

Hence, draw $\alpha\beta$ vertically upwards of length 155 units, and let the straight line drawn through α parallel to AB meet the horizontal through β in γ. Then $\beta\gamma$ represents $\frac{3}{4}T$.

On measuring $\beta\gamma$, we find that

$$\tfrac{3}{4}T = 116.25,$$

$$\therefore \quad T = 155 = S.$$

212   EQUILIBRIUM OF COPLANAR FORCES.

EXAMPLES XIII.

1. A uniform ladder, weighing 190 pounds, and of length 41 feet, rests with one end against a smooth vertical wall, and with the other end upon the ground ; if it is prevented from slipping by means of a peg at its lowest point, which is distant 9 feet from the wall, find the pressures on the peg, the ground, and the wall, when a man of 10 stone is standing on the ladder three-quarters of the way up.

2. A uniform ladder, weighing 120 pounds, and of length 34 feet, rests with one end against a smooth vertical wall, and with the other end upon the ground ; if it is prevented from slipping by means of a peg at its lowest point, distant 16 feet from the wall, find the reactions of the peg, the ground, and the wall, when a man weighing 150 pounds is standing two-thirds of the way up.

3. A uniform ladder is placed against a smooth vertical wall ; the bottom of the ladder is 6 feet from the wall, and the top 8 feet from the ground ; the mass of the ladder is 12 pounds, and a man of 10 stone stands on the ladder 2 feet from the bottom. Find the pressure of the ladder on the wall, and the reaction of the ground in direction and magnitude.

4. A uniform rod AB, of length one foot, and mass 8 pounds, is capable of turning freely in a vertical plane about a point O, distant 3 inches from A. The end B is loaded with 16 pounds, and the rod is kept in a horizontal position by a string AC, 5 inches long, attached to the end A, and to a fixed point C, situated vertically below O. Find the tension of the string, and the action at O in direction and magnitude.

5. A uniform rod AOB, of mass 5 pounds, is capable of turning freely about a fixed point O. The end B rests against a smooth vertical wall, and from the end A is suspended a mass of 7 pounds. Find the reaction of the wall at B, and also the action at the hinge O, supposing that $AO = \frac{1}{3}AB$, and that the distance of A from the wall is 4 times its distance above B.

CHAPTER XIV.

POLYGON OF FINE LIGHT RODS SMOOTHLY JOINTED AT THEIR EXTREMITIES.

120. *A number of fine light rods are freely jointed at their extremities to form a closed plane polygon ; the framework is in equilibrium under the influence of forces applied at the joints in the plane of the polygon. It is required to consider the conditions of equilibrium, and to determine the stresses in the rods.*

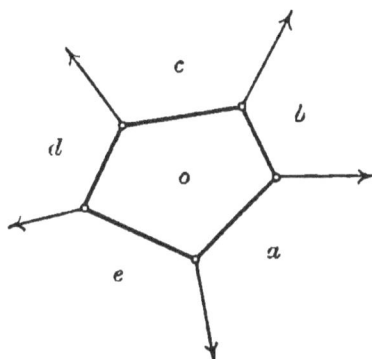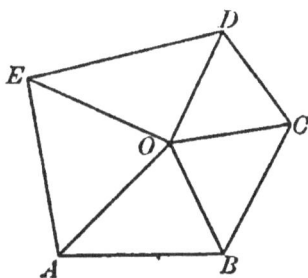

FIG. 130.   FIG. 130 *a*.

Let the figure on the left represent the polygon of rods, in equilibrium under the influence of forces applied at the joints in the directions indicated.

The framework of rods and the lines of action of the applied forces divide the plane, in which the space diagram is drawn, into a number of portions which are lettered o, a, b, c, d, e in the figure. The rod separating the space o from the space a is called the rod oa; the line separating the space a from the space b is called the line ab, and so on. The joint at the common extremity of the rods oa, ob is called the joint oab, and so on.

The whole framework must be in equilibrium considered as a rigid body. That is, if the joints were to become stiff so that they would not work, and if the same forces were applied as before, the polygon of rods would still be in equilibrium.

Hence, if straight lines AB, BC, CD, DE be drawn to represent the forces applied in the straight lines ab, bc, cd, de respectively, the straight line EA, which closes the force polygon, must represent the remaining force applied in the line ea.

We might now proceed to construct a funicular polygon corresponding to an arbitrarily chosen pole, but we shall have to consider the equilibrium of the different parts of the framework, and we shall see that the polygon of rods is itself a funicular polygon corresponding to a pole which we can find.

The consideration of the equilibrium of each rod separately, tells us that each rod is in a state of direct compression or tension.

Now consider the equilibrium of the portion of matter in the immediate neighbourhood of the joint oab. It is acted upon by three concurrent forces, namely, the force applied in the straight line ab, and the reactions of the adjoining portions of the rods oa, ob. Draw AO, BO parallel to the rods ao, bo respectively, to meet in O. Then $ABOA$ (this way round) is the triangle of forces for the joint abo.

Now consider the equilibrium at the joint bco. Since BO represents the action of the rod bo at the joint abo, OB must represent the action of the same rod at the joint bco. Also BC represents the force applied in the line bc. Therefore, joining OC, we see that $BCOB$ (this way round) is the triangle of forces for the joint bco. So that CO is parallel to the rod co, and represents its action upon the joint bco.

Proceeding in this way, and joining OD, OE, we see that OD, OE are parallel to the rods od, oe respectively and represent the stresses in those rods. Also, in considering the equilibrium of any joint, the direction arrows round the triangle determine whether the rods which meet at that joint are *struts* or *ties*. All *tie* rods may be replaced by strings.

We see that for equilibrium it is necessary and sufficient that the force polygon should close, and that the lines drawn from the angular points of the force polygon parallel to *corresponding* rods should be concurrent, these lines representing the stresses in the rods. The student should notice particularly the exact manner in which the diagrams *correspond*. The line representing the stress in any rod is drawn from the common extremity of the lines which represent the forces applied at the extremities of that rod. The polygon of rods is, in fact, a funicular polygon for the system of applied forces corresponding to the pole O.

121. As a typical example, let us consider the following:

The framework considered above rests in a given position of equilibrium. One of the applied forces is known completely, and the lines of action, but not the magnitudes, of all the others are known with one exception, the remaining force being wholly unknown. It is required to find the unknown forces and the stresses in the rods.

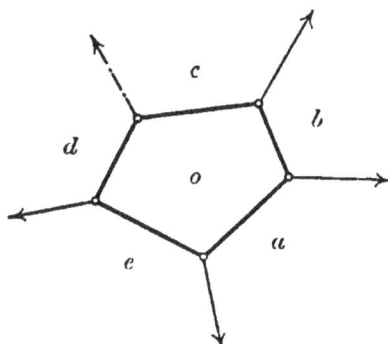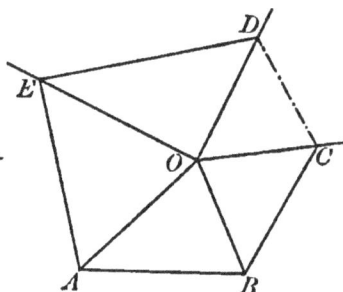

FIG. 131.   FIG. 131 a.

Let the lines *ab, bc, de, ea* be known, the dotted line *cd* being at present unknown. Then we can draw the space diagram with the exception of the dotted line.

Let the force applied in the line *ab* be known, the other forces being at present unknown.

To construct the force diagram, we take *AB* to. represent the given force applied in the line *ab*. Then we draw *AO, BO* parallel to *ao, bo* respectively, to meet in *O*. We are then able to draw the directions of *OC, OD, OE* parallel to *oc, od, oe* respectively. The point *C* is given by drawing *BC* parallel to *bc*, to meet *OC* in *C*. Similarly we get the point *E*, and then the point *D*. Joining *CD*, we have the magnitude and direction of the force applied at the joint *ocd*. The force diagram is now fully drawn, and the magnitudes of any of the unknown forces can be found by measuring the lines of the force diagram. The space diagram is completed by drawing *cd* parallel to *CD*.

122. The simplest example is that in which the framework is triangular. The figure below represents a triangular framework resting in a vertical plane on smooth horizontal supports at the joints *obc, oca*; the rod *oc* is horizontal, and a given vertical load is applied at the joint *oab.*

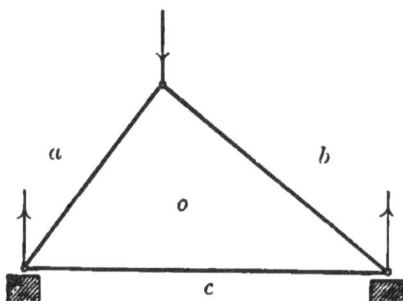
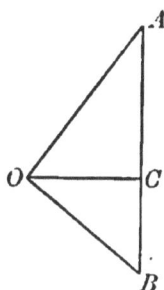

FIG. 132.   FIG. 132 a.

The force diagram is shown on the right. The rod *oc* is a *tie*, and the other two are *struts*.

It is often convenient to indicate in the space diagram which of the rods are ties, and which are struts. This we can do by drawing a double or thick line to indicate a strut, and a single or thin line to indicate a tie. Thus:

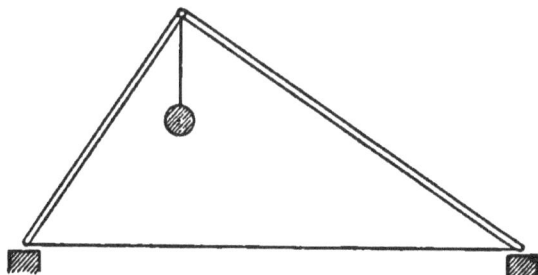

FIG. 133.

Or, it is found convenient to mark a strut +, and a tie −.

123. Ex. 1. *Four fine light rods, of lengths* 4, 3, 4, 3 *feet respectively, are freely jointed at their extremities to form a parallelogram.   To one of the angular points*

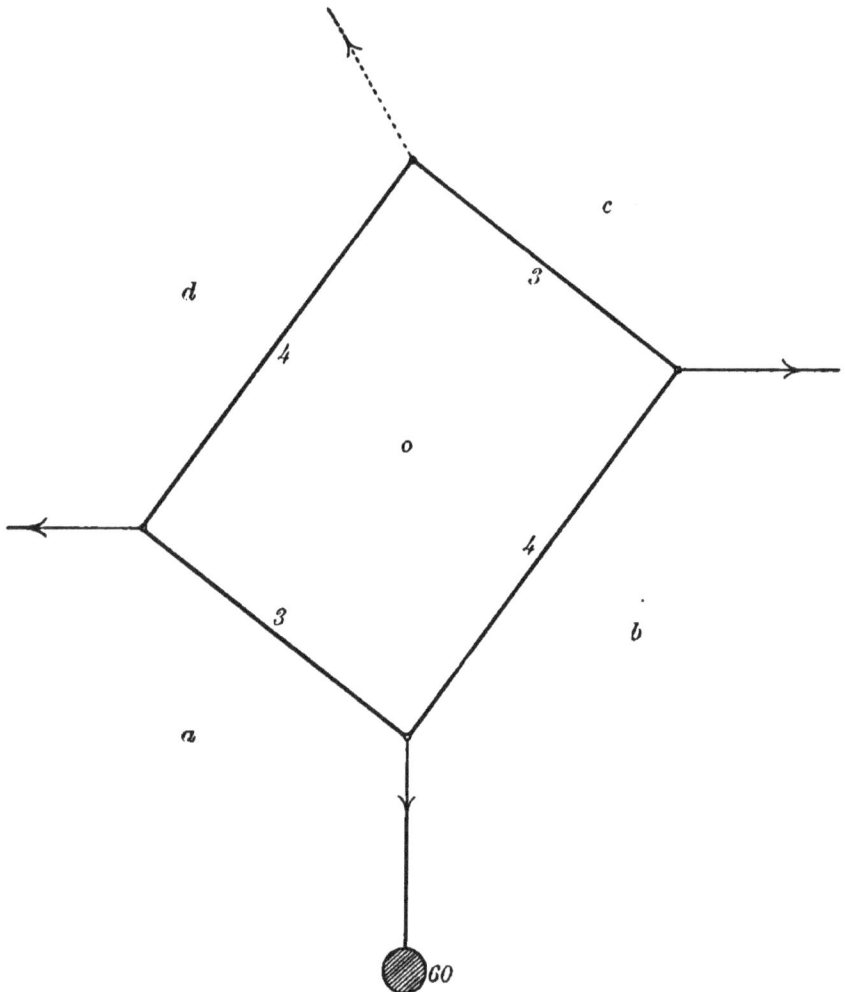

Fig. 134.

there is attached a mass of 60 *pounds, and the whole is supported at the opposite angular point.·   What horizontal forces must be applied at the other two angular points, in order that the framework may rest with the lowest joint 5 feet vertically below the point of support? Find also the stresses in the rods.*

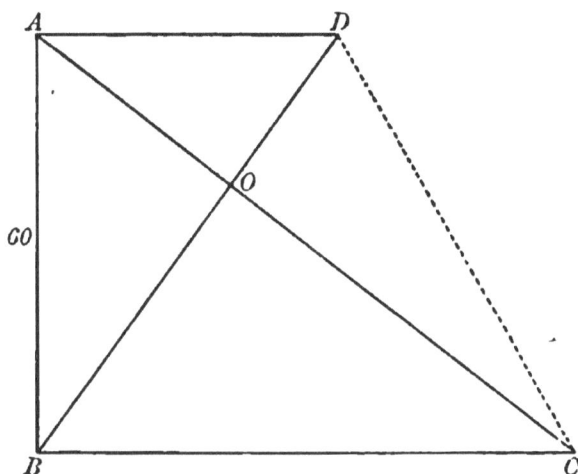

FIG. 134*a.*

Having constructed the space diagram to scale, let the different portions of the figure be marked with the letters *o, a, b, c, d,* as indicated. The line *cd*, being the line of action of the force of constraint at the point of support, is at present unknown.

With any suitable scale, draw *AB* vertically downwards, of length 60 units, to represent the tension of the string which supports the mass. Draw through *A* and *B* straight lines *AOC, BOD* parallel to *ao, bo* respectively, to meet the horizontals through *B* and *A* in *C* and *D* respectively.

On measuring the lines *OA, OB, OC, OD, BC, DA,* we find that the tensions of the rods *oa, ob, oc, od* and the two horizontal forces are 36, 48, 64, 27, 80, 45 pounds' weight respectively.

Also, *CD* determines the magnitude and direction of the force of constraint at the point of support.

124. *Ex. 2. The extremities H and K of a fine light rod, of given length, are connected with a fixed point L by two fine light strings, each of given length. From H and K are suspended two masses of given weights. It is required to find the position of equilibrium, the tension of each string, and the thrust in the rod.*

The data are sufficient to enable us to construct the shape of the framework, but not its position relatively to the vertical.

Let W_1 and W_2 be the measures of the loads applied at *H* and *K* respectively.

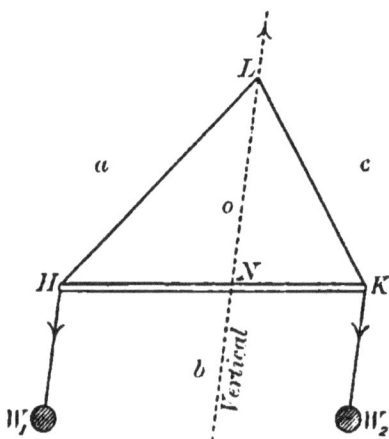

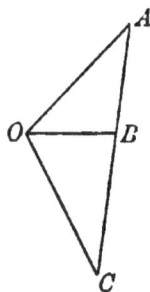

FIG. 135.   FIG. 135 a.

Considering the equilibrium of the triangle *HKL* as one rigid body, we see that the action at *L* has to balance two vertical forces W_1 and W_2 at H and K respectively. Hence we divide HK in the point N, so that $HN : NK = W_2 : W_1$. Then LN must be vertical. This determines the position of equilibrium, and the force diagram can now be constructed.

125. Ex. 3. *In the system indicated below, the point obc is fixed, and the joint oca is connected with another fixed point by a fine light rod ac. A given force is applied in a given direction ab to the joint oab. It is required to find the stresses in the rods.*

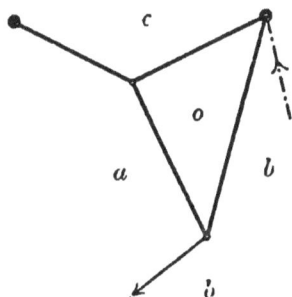

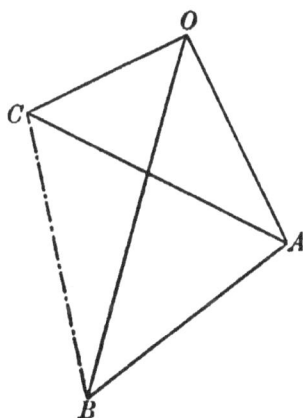

FIG. 136.   FIG. 136 a.

We can draw AB to represent the given force; then AO and BO, determining the point O; then OC and AC, determining the point C.

The straight line BC represents the constraint upon the hinge obc; the straight line CA represents the constraint upon the rod ac at its fixed extremity.

126. Ex. 4. *The accompanying figure represents a framework of four light rods freely jointed at their* *extremities. The points obc and oda are fixed, and forces P and Q of given magnitudes are applied in given directions at the other two joints, as indicated. It is required to find the stresses in the rods, and the actions at the two fixed points.*

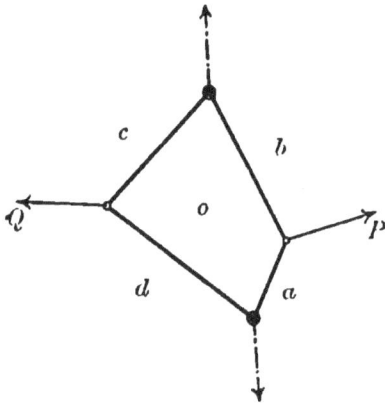

FIG. 137.

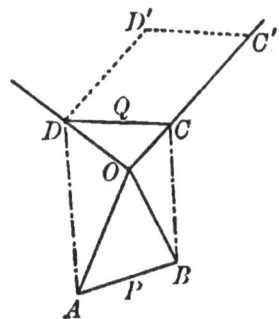

FIG. 137 a.

Draw AB to represent P (which acts along the line ab) in magnitude and direction. Draw AO, BO parallel to ao, bo respectively. This gives the point O. The directions of OC, OD can be drawn parallel to oc, od respectively.

Take any point C' in OC, and draw $C'D'$ to represent Q (which acts along cd) in magnitude and direction. Draw $D'D$ parallel to CO, to meet OD in D. Through D draw DC parallel to $D'C'$, to meet OC in C. Thus we have the points C and D. We complete the force diagram by joining B, C and D, A. Then BC and DA represent the actions at obc and oda respectively, and the stresses in the rods are given by OA, OB, OC, OD.

Below we give the figures for the case in which
P and Q are parallel, and in the same direction.

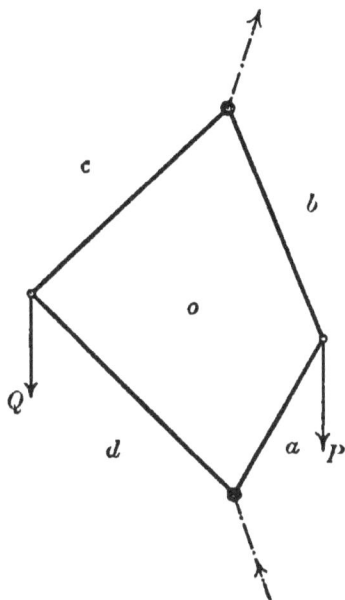

FIG. 138.

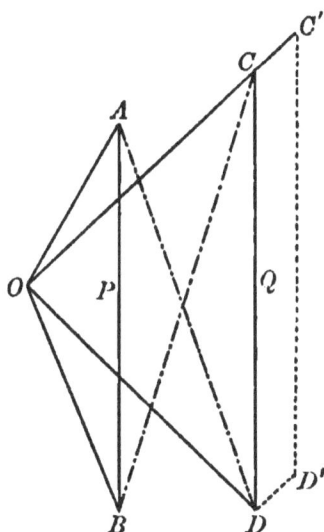

FIG. 138 a.

EXAMPLES XIV.

1. Two fine light rods AB and BC, each of length 5 feet, are
freely jointed at B, and rest in a vertical plane, upon a smooth
horizontal plane at A and C. A load of 40 pounds is applied
at B, and the system is kept from collapsing by a fine light
string, of length 6 feet, connecting the extremities A and C of
the rods. Find the reactions at A and C, the tension of the
string, and the thrusts in the rods.

2. The extremities H and K of a fine light rod are connected
with a fixed point L by two fine light strings, each equal in
length to the rod. From H and K are suspended masses of 21 and
35 pounds respectively. Find the position of equilibrium, the
tensions of the strings, and the thrust in the rod.

3. Three fine light rods BC, CA, AB, of lengths 15, 13, 14 inches respectively, are freely jointed at their extremities to form the triangular framework ABC, which is capable of turning freely about the fixed point A. To the joint C there is applied a force of 56 pounds' weight in a direction perpendicular to AB, and outwards from the triangle. What force must be applied at the joint B, in a direction perpendicular to the rod CA, to preserve equilibrium? Find also the stresses in the rods and the action at the hinge A.

4. Three equal rods, of no appreciable weight, are freely jointed together at their extremities to form a triangular framework ABC, which is capable of turning freely in a vertical plane about the joint A, which is fixed. A mass of 100 pounds is suspended from B, and the framework is sustained in a position in which AB is horizontal, and C uppermost, by means of a horizontal fine string CD, which connects C with a fixed point D. Find the tension of the string and the stresses in the rods; determine also the magnitude and direction of the reaction at A.

5. The triangular framework of Question 3 is capable of turning freely about the joint B, which is fixed; a load of 168 pounds is applied at C, and the whole is supported with AB horizontal, and C below AB, by means of a vertical force applied at A. Find the magnitude of this force, the reaction at B, and the stresses in the rods.

6. Four fine light rods are freely jointed at their extremities to form a quadrilateral $ABCD$, the rods AB, AD being each of length 7·5 inches, and the rods BC, CD each of length 11·7 inches. A mass of 48 pounds is attached at C, and the whole is supported at A. What horizontal forces must be applied at B and D so that the points B and D may rest 9 inches apart in a horizontal line? Find also the tensions of the rods.

7. Four fine rods, of no appreciable weight, and each of length 5 feet, are freely jointed at their extremities to form a rhombus $ABCD$, which is placed between two parallel walls, distant 8 feet apart, so that the framework touches the walls at B and D, and the straight line BD is horizontal and perpendicular to the walls.

If the joints A and C are pressed towards one another, with forces each of 60 pounds' weight, find the pressures upon the walls.

8. In Art. 123, suppose that the line da passes through the opposite angular point, and that otherwise the data are unaltered. Find the forces applied in the lines bc, da, and the stresses in the rods.

9. Four fine light rods are freely jointed together at their extremities to form a parallelogram $ABCD$, AD being of length 9 inches and AB of length 13 inches. The point A is fixed, and C is attached to a fixed point E, by a fine elastic string, so that the points A, C, E are in one straight line. The points B and D are pulled apart 10 inches by forces of 100 pounds' weight, applied at B and D in the line BD. Find the tension of the string and the actions in the rods.

10. A fine light string $ABCDEF$, of length 33 inches, has its extremities fixed at two points A and F, situated 27 inches apart in a horizontal line. The portions of string AB, BC, CD, DE, EF are of lengths $8\frac{1}{2}$, 5, 6, 5, $8\frac{1}{2}$ inches respectively, and another light string, of length 12 inches, connects the points B and E. Two masses, each of 24 pounds, are suspended from C and D, and the whole system rests in a symmetrical position with BE and CD horizontal. Find the tension of each portion of string.

11. A light rod AB, of length 1 foot, rests in a horizontal position, with masses each of 16 pounds suspended from A and B. A fine light string $ACDB$, of length 16 inches, has its extremities attached at A and B, and the whole is supported by means of two vertical forces applied at C and D. The portions of string AC, CD, DB are of lengths 5, 6, 5 inches respectively, and the whole rests in a symmetrical position with CD horizontal. Find the tension of each portion of string.

12. Four fine light rods, each of length 5 feet, are freely jointed at their extremities to form the rhombus $ABCD$. The joints B and D are fixed, B being 6 feet vertically above D, and masses of 30 and 60 pounds are hung from A and C respectively. Find the stresses in the rods and the magnitudes of the reactions at B and D.

D.S.   P

13. Two fine light rods AB, BC, each of length 2 feet, are freely jointed together at the point B, which is fixed. The rods rest in a horizontal position, with masses of 70 and 84 pounds hung from A and C respectively, being supported by two fine strings connecting A and C with a fixed point D, situated 7 inches vertically above B. Find the magnitudes of the actions at B and D, the tensions of the strings, and the stresses in the rods.

14. A fine light string $ABCDE$, of length 49 inches, has its extremities attached to the fixed points A and E, situated $33\cdot8$ inches apart in a horizontal line, and supports at its middle point C a mass of 78 pounds. Two fine light rods BF, FD, each of length $15\cdot6$ inches, are capable of turning freely about the fixed point F, which is situated at the middle point of AE; the rods are attached to the string at points B and D, distant $6\cdot5$ inches from A and $6\cdot5$ inches from E respectively. Find the tensions of the different parts of the string, and the thrusts in the rods.

15. Three fine light rods are jointed together at their extremities to form a triangle ABC, which is right-angled at B. The framework rests in a vertical plane upon a smooth horizontal plane at A and C, and supports a load at B. If BD is drawn perpendicular on AC, prove that the reactions at A and C, the load at B, the thrusts in the rods AB, BC, and the tension of the rod AC are proportional to DC, AD, CA, BC, AB, BD respectively.

16. Four fine light rods are freely jointed at their extremities to form a parallelogram $ABCD$, which is in equilibrium under the influence of forces P, Q, R, S, acting at A, B, C, D respectively. If the forces P and R are equal and opposite, their lines of action passing through the middle points of BC, DA respectively, prove that the forces Q and S are also equal and opposite, and that their lines of action pass through the middle points of DA, BC respectively. Prove also that, if H be the middle point of BC, the forces P and Q, and the tensions of the rods AB, BC, CD, DA, are proportional to HA, DH, AB, BH, CD, HC respectively.

17. Four fine light rods are freely jointed at their extremities to form a quadrilateral framework $ABCD$, which is in equilibrium

under the influence of forces applied at A, B, C, D. If the forces applied at A and B are equal, and in opposite directions, prove that BC must be parallel to AD, and that the forces at C and D are also equal and in opposite directions.

18. Four fine light rods are freely jointed at their extremities to form a quadrilateral framework $ABCD$, in which AD is parallel to BC, and the framework is kept in equilibrium under the influence of forces P, P, Q, Q, applied at A, B, C, D, in directions CD, DC, AB, BA respectively. Prove that the forces P and Q, and the stresses in the rods AB, BC, CD, DA, are proportional to CD, AB, AB, $BC \sim AD$, CD, $BC \sim AD$ respectively.

19. Three fine light rods are freely jointed at their extremities to form a triangular framework ABC, which is in equilibrium under the influence of three forces applied at A, B, C, in the lines OA, OB, OC respectively. Any point A' is taken in BC, and straight lines through A', parallel to BO and CO, meet AO in C' and B' respectively. Prove that

(i.) The straight lines through C' and B', parallel to AB and CA respectively, meet at a point O' in BC.

(ii.) The forces in the lines OA, OB, OC, and the stresses in the rods BC, CA, AB, are proportional to $B'C''$, $C'A'$, $A'B'$, $O'A'$, $O'B'$, $O'C''$ respectively.

(iii.) If O is the centre of a circle which touches each side of the triangle ABC, O' is the centre of the circle which circumscribes the triangle $A'B'C''$, and hence the stresses in the three rods are equal to one another.

(iv.) If O is the orthocentre of the triangle ABC, the figure $A'B'C'O'$ is similar to the figure $ABCO$, and hence the forces in the lines OA, OB, OC, and the stresses in the rods BC, CA, AB are proportional to BC, CA, AB, OA, OB, OC respectively.

(v.) If O is any point on the circle which circumscribes the triangle ABC, the figure $A'B'C'O'$ is similar to the figure $ABCO$, and hence as in (iv.).

(vi.) If O is the intersection of the medians of the triangle ABC, O' is the intersection of the medians of the triangle $A'B'C'$; also the forces in the lines OA, OB, OC, and the stresses in the rods BC, CA, AB are proportional to $3OA$, $3OB$, $3OC$, BC, CA, AB respectively.

(vii.) If $ABOC$ is a parallelogram, $C'A'B'O'$ is similar to $OBAC$, and hence the forces in the lines OA, OB, OC, and the stresses in the rods BC, CA, AB are proportional to OA, OB, AB, BC, CA, OC respectively.

20. A number of fine light rods are jointed together at their extremities to form a closed plane polygon. O is any point in the plane of the polygon. Show that a system of forces can be found such that, acting at the joints along straight lines which intersect at O, they will keep the framework in equilibrium. Prove also that the stresses in the different rods are inversely proportional to the lengths of the perpendiculars from O upon those rods respectively.

21. Four fine light rods are freely jointed at their extremities to form a parallelogram $ABCD$. O is a point which divides AC in the ratio of $2 : 1$. The framework is in equilibrium under the influence of forces applied at the joints A, B, C, D, in the directions OA, OB, OC, OD respectively. Prove that the forces in the lines OA, OB, OC, OD, and the tensions of the rods AB, BC, CD, DA, are respectively proportional to $3OA$, $6OB$, $12OC$, $6OD$, $2AB$, $4BC$, $4CD$, $2DA$.

22. A number of fine light rods are jointed freely at their extremities to form a closed polygon $ABCD...$, which is in equilibrium under the influence of forces acting at the joints A, B, C, D, ..., in the lines OA, OB, OC, OD, ... respectively. Prove that, if the stresses in the rods AB, BC, CD, ..., and the forces at the joints A, B, C, D, ... are proportional to AB, BC, CD, ..., OA, OB, OC, OD, ... respectively, (i.) the number of rods is six, (ii.) opposite rods are equal to one another, (iii.) each diagonal is parallel to each of two opposite sides and equal to the sum of those sides, (iv.) the point O is the middle point of each diagonal.

CHAPTER XV.

OPEN POLYGON OF FINE LIGHT RODS.
SUSPENSION BRIDGE.

127. Let us suppose that in a framework, like that considered in the last chapter, one of the rods is missing, and that the two extreme rods are freely hinged at their extremities to two fixed points. The figure below indicates such a system of rods, the extreme rods oa and oe being hinged to the fixed points H and K respectively.

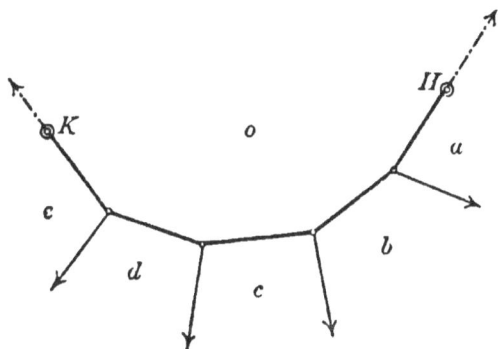
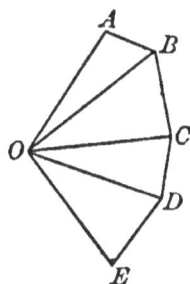

FIG. 139.   FIG. 139 a.

The external forces acting upon the rods oa and oe, at their extremities H and K respectively, are evidently in the directions of the rods themselves.

The student will have no difficulty in constructing the force diagram, which is done just as in the preceding chapter. The polygon $OABCDEO$ is the force polygon for the whole system. For the joint abo we have the triangle $ABOA$; for the joint bco the triangle $OBCO$, etc.

I. If the rods are all given in position, and also the directions of the applied forces, we can draw the force diagram, provided we know the magnitude of one only of the applied forces.

Suppose, for instance, that all the lines of the space diagram are given, and that the force which acts along the line bc is fully known. We can then draw BC and the straight lines BO, CO meeting in O. Then the directions of the lines OA, OD, OE can all be drawn. We can also draw the directions of BA, CD, thus determining the points A and D respectively; and from D we can draw DE.

II. If the applied forces are all fully known, that is, in magnitude, direction, and line of action, we can complete both diagrams, provided we know the positions of two adjoining rods only.

For, as the forces are all known, we can construct the line $ABCDE$. If also the rods ob, oc are given in position, we can draw OB, OC, thus determining the point O; and the rest of both figures is easily completed. This will not, however, give us the positions of the points H and K, unless the lengths of the extreme rods are given.

128. *A fine light string is attached at its extremities to two fixed points, and rests in a vertical plane under loads applied to it at different points.*

This is a particular case of the preceding.   The line
ABCDE is now straight and vertically downwards.

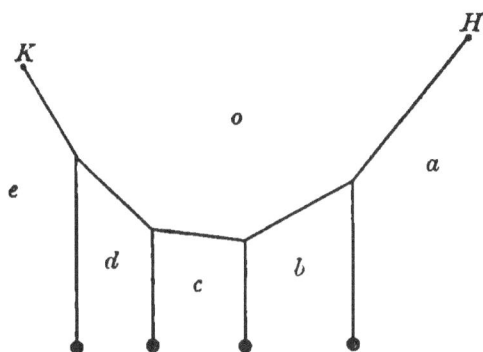 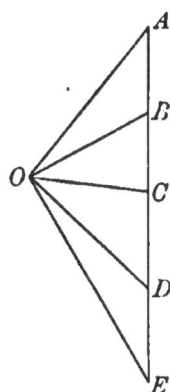

FIG. 140.   FIG. 140*a*.

If we imagine the different portions of the string
to be replaced by bars, and the whole polygon inverted
and changed from left to right, the shape remaining
unaltered, the same force diagram would apply.

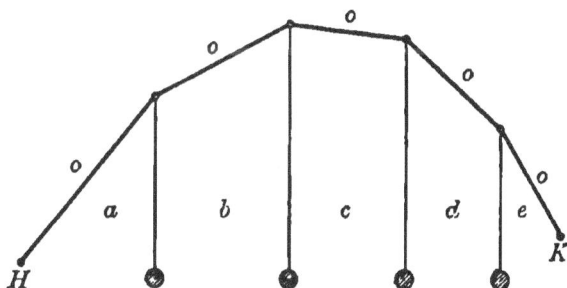

FIG. 141.

In this case, however, the slightest disturbance would
cause the whole thing to collapse.   The equilibrium is
what is called *unstable*.   See Art. 41.

129. The case in which a fine light string rests
under loads, which are all equal and at equal horizontal
distances apart, is important.

The student is recommended to draw the figures for himself. He will take a number of equidistant parallel vertical lines, marking the spaces between them a, b, c, d, etc. Then he will draw the straight line $ABCD\ldots$ vertically downwards, marking off AB, BC, CD, etc.,

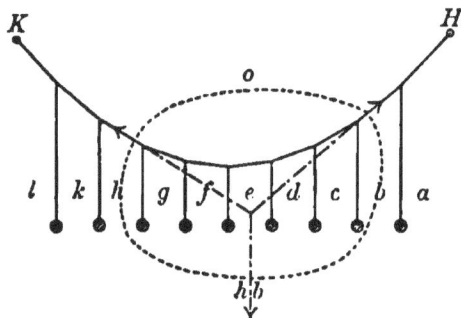

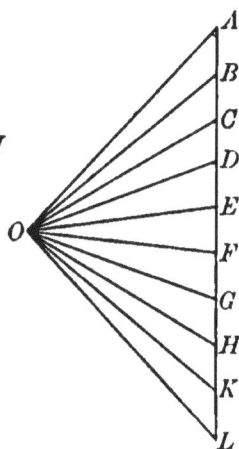

FIG. 142.   FIG. 142 a.

all equal to one another. He will draw two straight lines od, oe across the spaces marked d and e respectively, and OD, OE parallel to od, oe respectively. This gives the point O, and the rest easily follows.

An examination of the two figures will show at once that the tension of each portion of the string is proportional to its length.

130. We have drawn in the space diagram of the preceding article a closed curve intersecting two of the strings ob, oh, and enclosing a portion of the system within it.

Regarding the matter included within this curve as one rigid body, let us consider the equilibrium of the forces acting externally upon it. These are: The tensions of the two strings ob, oh, together with the weights of all the parts included. The weights of all these parts reduce to a single force represented by BH, and acting along a vertical straight line situated *midway* between bc and gh.

∴ *the strings* ob, oh *intersect at a point midway between* bc *and* gh, *and the triangle of forces for this portion of the system is* $OBHO$.

This gives a method of constructing the space diagram without drawing the force diagram:

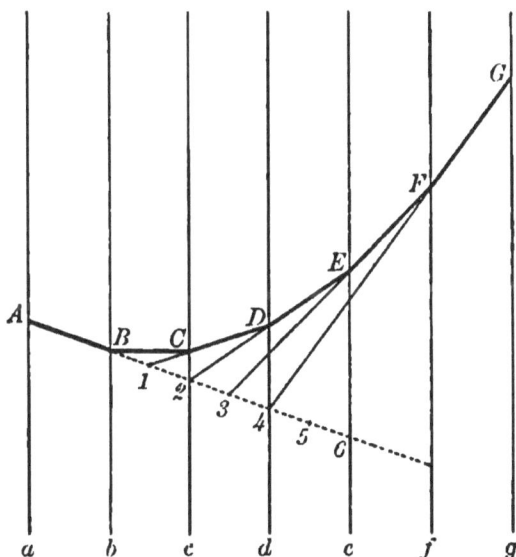

FIG. 143.

Let Aa, Bb, Cc, etc., be consecutive parallel lines in which the loads are applied, and let the strings be denoted by AB, BC, CD, etc. We will suppose that the two strings AB, BC are given.

Produce the line AB to meet Cc, Dd, Ee, etc., in the points *2*, *4*, *6*, etc., and find the middle points *1*, *3*, *5*, etc., of *B2*, *24*, *46*, etc.

Then, joining *1C*, and producing to D, we have the string CD; also *2D* gives the string DE, *3E* gives EF, and so on.

131. Now let us suppose that the loads are sufficiently numerous to enable us to look upon the string as a continuous curve instead of a series of straight lines. The above piece of work will enable us to examine the nature of this curve.

We will suppose that the points H and K are situated in a horizontal line. Let P and P' be any two points on the curve, and let the tangents at P and P' intersect in Q. Draw PN, QR, $P'N'$ perpendiculars to HK.

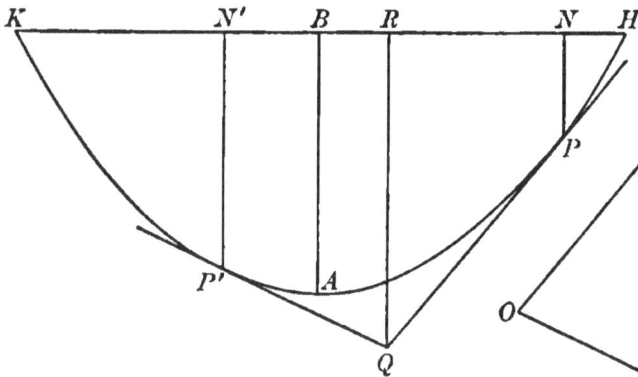
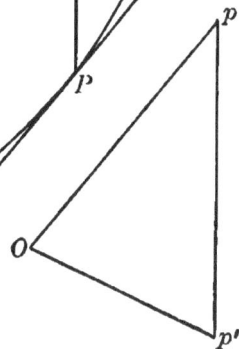

FIG. 144.   FIG. 144 a.

Then the curve must be such that the point Q lies midway between PN and $P'N'$ for all positions of P and P'; that is, R must be the middle point of NN'. Also, the force diagram for the portion of string between P and P' is a triangle $pp'Op$, in which pp' represents the total weight supported between P and P', and is therefore proportional to NN', and $p'O$, Op are parallel to the tangents at P' and P respectively, and represent the tensions at those points.

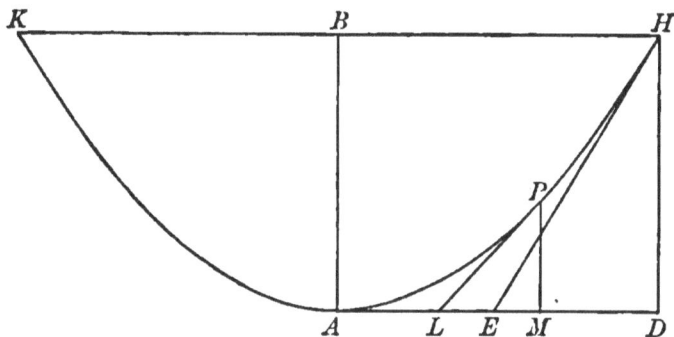

FIG. 145.

If P and P' be taken to coincide with H and K respectively, Q will lie vertically below B, the middle point of HK, and, as H and K may be taken to be any two points of the string which lie in a horizontal line, it follows that the figure is symmetrical about the vertical through B. Let A be the lowest point of the curve, then A is vertically below B.

Take P' to coincide with A, and let the tangent at P and the vertical through P meet the horizontal through A in L and M respectively. Then L is the middle point of AM,

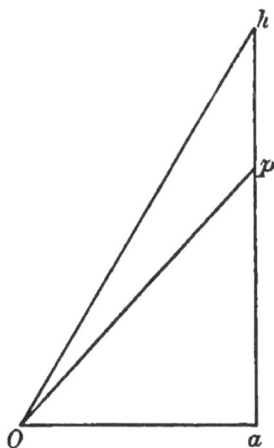

FIG. 145 a.

and the triangle of forces for the portion AP is $paOp$, where aO is horizontal, and represents the tension at A, Op is parallel to LP, and represents the tension at

P, and pa represents the whole weight supported between A and P, and is proportional to AM.

Now let P move up to H, and let L, M, p become E, D, h respectively for this new position of P. Then E is the middle point of AD, and ha represents half of the whole weight supported by the string. Thus

$$ap : ah = AM : AD$$

[We have

$$\frac{PM}{AM} = \frac{1}{2} \cdot \frac{PM}{LM} = \frac{1}{2} \cdot \frac{ap}{Oa} = \frac{1}{2} \cdot \frac{ap}{ah} \cdot \frac{ah}{Oa} = \frac{1}{2} \cdot \frac{AM}{AD} \cdot \frac{DH}{ED}$$

$$= \frac{AM}{BH} \cdot \frac{AB}{BH};$$

$$\therefore \quad \frac{PM}{AB} = \left(\frac{AM}{BH}\right)^2.$$ That is, PM varies as the square of AM.

The student of Higher Mathematics will know that this shows that the curve is a *parabola*, with its axis vertically upwards.]

132. If we know the positions of the points H, K, A and the whole weight supported, we can determine (i.) the directions of the string at H and K, and the tensions at those points and at A ; (ii.) the tension at any point of the string where its direction is given; (iii.) the tension at any given horizontal distance from A ; and (iv.) the position of any number of points on the string.

(i.) The positions of the points H, K, A determine the rectangle $ADHB$. We can therefore find E, the middle point of AD, and, joining EH, we have at once the direction of the string at H.

If the whole weight supported is known, we can draw ha vertically downwards, and of such a length that it represents half the weight. Then we draw hO parallel to HE, to meet the horizontal through a in O. Measuring Oh and aO, we have the tensions at H and A respectively.

(ii.) Draw Op in the given direction, to meet ah in p. Then Op represents the tension at that point of the string where it is parallel to Op.

(iii.) Take AM equal to the given horizontal distance, and find a point p in ah such that $ap : ah = AM : AD$. Then Op represents the tension at the point P, which is vertically over M.

(iv.) We can obtain the position of the point P, corresponding to a given position of M, by bisecting AM in L, and drawing LP parallel to Op, to meet the vertical through M in P. By varying the position of M we can get as many points as may be required.

133. The problem we have been discussing is approximately that of the ordinary *suspension bridge*. Here we may neglect the weight of the chain, and that of the suspending rods, in comparison with that of the roadway. The loads suspended from successive portions of the chain are equal portions of the roadway. If the lengths of the pieces are so adjusted that the curve of the string takes up the shape we have been considering, the tensions of the supporting rods would not then tend to break or bend the roadway, which must be made strong enough to bear without bending, the strain due to loads moving across it.

134. Ex. 1. *The figure below represents a beam, of length* 50 *feet, and weighing* 100 *pounds, resting in a horizontal position, being supported by five light vertical strings connecting it with a light chain. The*

strings are at equal horizontal distances apart of 10
*feet, the middle one being attached to the middle point
of the beam. The lowest, parts of the chain are each
inclined at an angle of* 85° *to the vertical. Determine
the directions of the remaining parts of the chain, and
the tensions of those parts, in order that the tensions
of the five strings may be equal to one another.*

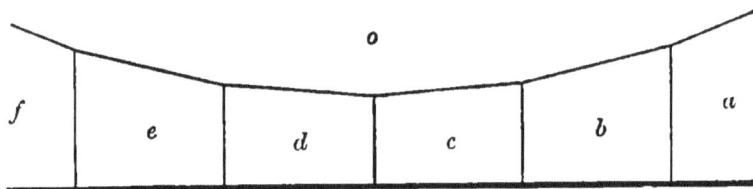

FIG. 146.

We cannot at the outset draw the whole chain.   Take
a straight line *ABCDEF* vertically downwards, of length

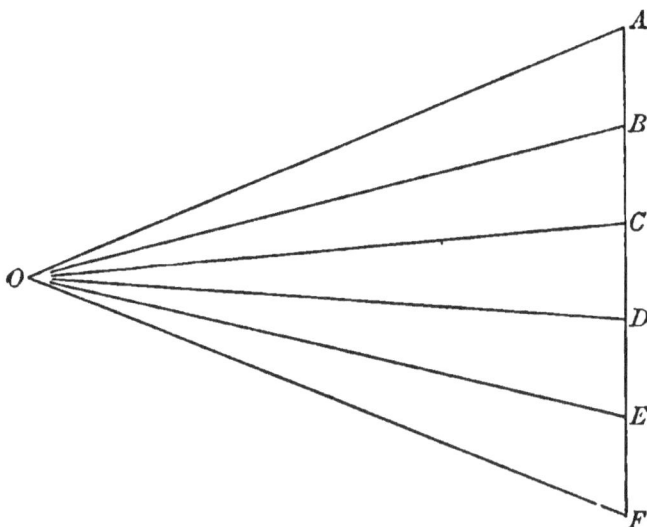

FIG. 146 *a*.

100 units, and make $AB = BC = CD = DE = EF = 20$
units.

Make the angles DCO, CDO each equal to 85°, thus determining the point O, and draw OA, OB, OE, OF.

Then, on measuring the lines OA, OB, OC, we see that

the tensions of oc, od are each 115 pounds' weight,

       ,, ob, oe ,,   118   ,,

       ,, oa, of ,,   125   ,,

Also the angles OBE, OEB are found to be each $75\frac{1}{4}°$, and the angles OAF, OFA each $66\frac{1}{2}°$. Hence the strings ob, oe are each inclined at $75\frac{1}{4}°$ to the vertical, and the strings oa, of each at $66\frac{1}{2}°$.

135. Ex. 2. *A suspension bridge, 20 feet broad, and of 120 feet span, is supported by two parallel chains, each of which dips down 16 feet in the middle. The*

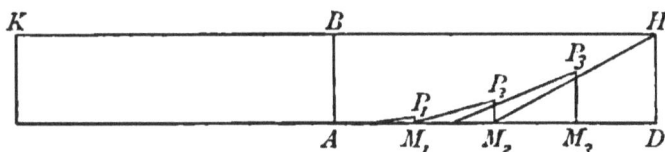

FIG. 147.

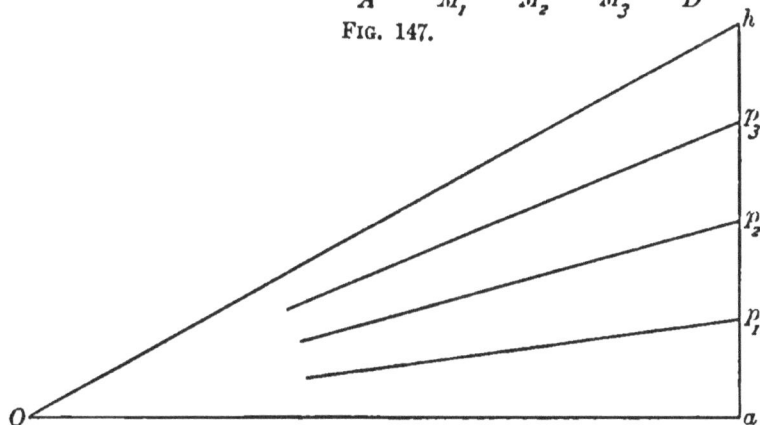

FIG. 147 a.

mass of the roadway is 100 pounds per square foot; find the tension of each chain at the lowest point, and.

also at points a quarter of the way across the bridge. If the shortest distance between the chains and the roadway is 2 feet, find the distances of the roadway below the chains at intervals of 15 feet, commencing at the centre of the bridge.

The total mass of the roadway = 120 × 20 × 100 pounds, therefore the total load upon each chain = 120,000 pounds.

Draw a horizontal straight line HBK, of length 120 units, to represent the span of the bridge. Through B, the middle point of HK, draw BA vertically downwards, of length 16 units, and let the horizontal through A meet the vertical through H in D. The point A represents the lowest point of the chain.

In AD take points M_1, M_2, M_3 so that

$$AM_1 = M_1M_2 = M_2M_3 = M_3D,$$

each portion representing 15 feet.

Draw ha vertically downwards, of length 60,000 units, to represent half the load supported by each chain, and let the horizontal through a meet in O the straight line drawn through h parallel to HM_2. Then Oa represents the tension of each chain at its lowest point.

In ah take points p_1, p_2, p_3 so that

$$ap_1 = p_1p_2 = p_2p_3 = p_3h.$$

Then Op_2 represents the tension of each chain at points a quarter of the way across the bridge.

Let the straight line through the middle point of AM_1, parallel to Op_1, meet the vertical through M_1 in P_1, also the straight line through M_1, parallel to Op_2, meet the vertical through M_2 in P_2, and the straight line through the middle point of M_1M_2, parallel to Op_3, meet the vertical through M_3 in P_3. Then the points

P_1, P_2, P_3 represent points on the curve formed by one of the chains.

On measurement we find that

$$Oa = 113,000, \quad Op_2 = 116,000.$$

Therefore the tension at the lowest point of each chain is 113,000 pounds' weight, and at a point a quarter of the way across the bridge 116,000 pounds' weight.

Also we find

$$P_1 M_1 = 1,$$

$$P_2 M_2 = 4,$$

$$P_3 M_3 = 9,$$

and, as the roadway is 2 feet below A, we see that the distances of the roadway below the chains at intervals of 15 feet, commencing at the centre of the bridge, are 2, 3, 6, 11, 18 feet.

EXAMPLES XV.

1. The light rods AB, BC, CD are freely jointed at B and C, and the points A and D are smoothly hinged to two fixed points situated in a horizontal line. If the figure $ABCD$ is one half of a regular hexagon when two masses are suspended from B and C, prove that the weights of the two masses are equal, and determine the stresses in the rods in terms of the weight of either mass.

2. Three light rods AB, BC, CD, of lengths 8, 5, 5 inches respectively, are freely jointed at B and C, and the points A and D are smoothly hinged to two fixed points, D being situated 7 inches vertically above A. A mass of 7 pounds is suspended from B, and the system rests with AB parallel to DC, being supported by a force applied at C in a direction opposite to the bisector of the angle BCD. Find the magnitude of this force and the stresses in the rods.

D.S.   Q

3. A light rod AB, of length 17 inches, is capable of turning freely about the end A which is fixed. A fine string, of length 28 inches, has one extremity attached at B, and the other extremity at D, a fixed point situated 3 inches vertically above A. Find what vertical force must be applied at C, a point of the string 15 inches from D, in order that the rod may rest in a horizontal position, with a mass of 48 pounds attached at B. Find also the tensions of both parts of the string and the stress in the rod.

4. A light rod AB is capable of turning freely in a vertical plane about the fixed point A. It is supported in a horizontal position, with a mass of 30 pounds attached at B, by means of a fine string BCD attached to a point D vertically above A. Find what force must be applied at the point C, in the direction AC, to preserve equilibrium, supposing that ACB is an equilateral triangle, and DCB a right angle. Find also the tensions of both parts of the string and the stress in the rod.

5. A fine light string $ABCD$ has its extremities fixed at A and D, and supports a mass of 100 pounds at B. What load must be applied at C, in order that BC may be horizontal, and the angles ABC, BCD equal to 120° and 150° respectively? Find also the tensions in the different parts of the string.

6. A beam weighing 100 pounds is supported in a horizontal position by means of four light vertical strings, arranged at intervals of 10 feet, which connect it with a light chain supported at its extremities. The extreme parts of the chain are inclined at angles of 40° to the vertical; determine (i.) the directions, (ii.) the lengths, (iii.) the tensions, of the other parts of the chain, in order that the tensions of the vertical strings may be equal to one another.

7. A suspension bridge, of 60 feet span, is 10 feet broad, and is supported by two parallel chains, each of which dips down 32 feet in the middle. The mass of the roadway is 40 pounds per square foot, and the shortest distance between the roadway and either chain is 1 foot. Find (i.) the inclination of the chain to the vertical at each end of the bridge, (ii.) the tension at each end, (iii.) the tension at the lowest point, (iv.) the distance of the

roadway below the chain at a point one-third of the way across the bridge.

8. A fine light string $ABCD$ has its extremities A and D fixed, and is in equilibrium under the influence of forces acting at B and C in directions DB, AC respectively. If the figure $ABCD$ is a parallelogram, prove that the tensions of AB, BC, CD and the forces acting at B and C are proportional to AB, BC, CD, DB, AC respectively.

9. A fine light string $HPAK$ has its extremities fixed at two points H and K, situated in a horizontal line, and rests in a vertical plane under the influence of a load distributed uniformly in a horizontal direction. A is the lowest point and P any other point of the string in the position of equilibrium. The horizontal through A meets the verticals through P and H in M and D respectively, and L is the middle point of AM. A straight line through M perpendicular to LP meets the vertical through A in O. Prove that, for all positions of P,

(i.) O is a fixed point.

(ii.) The tension at P, the tension at A, and the weight of the whole load supported are proportional to OM, OA, $2AD$ respectively.

(iii.) If S is the middle point of OA, $SP = MP + AS$.

(iv.) LP bisects the angle SPM.

(v.) If OM and SP intersect at N, $PM = PN$, and the locus of N is a circle with its centre at S.

(vi.) $AM^2 = 2OA \cdot PM$.

(vii.) If $AM = n \cdot AD$, n being a numerical fraction, $PM = n^2 \cdot HD$.

(viii.) If a straight line through A, perpendicular to OM, meets MP in Q, $QM = 2 \cdot PM$.

(ix.) Without making use of the point O, let a straight line through L, perpendicular to LP, meet the vertical through A in S. Then S is a fixed point, and the tension at P, the tension at A, and the weight of the whole load supported are proportional to SL, SA, AD respectively ; also S is the same point as before.

(x.) If $AD=2.HD$, the tension at A is half the weight of the whole load supported, and the string at H makes an angle of 45° with the vertical.

(xi.) At a point where the tension of the string is double the tension at the lowest point, the string makes an angle of 60° with the horizontal.

CHAPTER XVI.

STIFF QUADRILATERAL FRAMEWORK OF FINE LIGHT RODS.

136. *Four fine light rods are freely jointed at their extremities to form a quadrilateral, which is stiffened by another fine light rod connecting two opposite joints.*

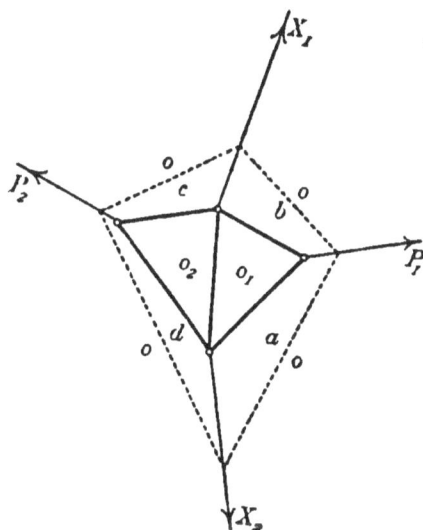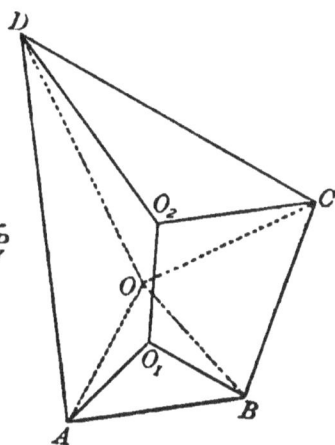

FIG. 148.   FIG. 148 a.

It is required to consider the equilibrium of the framework under the influence of forces applied at the joints.

Let P_1, X_1, P_2, X_2 be the measures of the forces applied at the joints abo_1, o_1bco_2, o_2cd, o_2dao_1, as indicated in the left-hand figure.

The force diagram for the point abo_1 will be a triangle ABO_1A (this way round), in which AB represents P_1 and AO_1, BO_1 are parallel to the rods ao_1, bo_1 respectively.

Now let BC be drawn representing X_1; then O_1B, BC represent two of the forces acting upon the joint o_1bco_2. The force diagram for the point o_1bco_2 is therefore the quadrilateral $O_1BCO_2O_1$ (this way round), in which CO_2, O_1O_2 are drawn parallel to the rods co_2, o_1o_2 respectively.

If now CD be drawn to represent P_2, then O_2C, CD represent two of the forces acting upon the joint o_2cd. Therefore DO_2 must represent the remaining one, i.e. DO_2 must be parallel to the rod do_2 and represent the action of that rod upon the joint o_2cd. The triangle of forces for the joint o_2cd is thus O_2CDO_2 (this way round).

Now, considering the joint ao_1o_2d, we see that three of the forces acting upon it are represented by AO_1, O_1O_2, O_2D. Hence the remaining one X_2 must be represented by DA, and the force diagram for the joint ao_1o_2d is the quadrilateral AO_1O_2DA (this way round).

Thus we see that for equilibrium it is necessary and sufficient that the force polygon representing the external forces applied to the framework should close, and that the line O_1O_2 should be parallel to the rod o_1o_2, where O_1 is found by drawing AO_1, BO_1 parallel to the rods ao_1, bo_1 respectively, and O_2 is found by drawing CO_2, DO_2 parallel to the rods co_2, do_2 respectively. When the figure is in this way completed the lines drawn parallel to corresponding rods represent the stresses in the rods.

137. It will be noticed that the framework in this case is not deformable, and is practically a rigid body. As the stresses in the rods can be of any magnitude and in either direction, it is necessary and sufficient for equilibrium that the four forces P_1, X_1, P_2, X_2 should form a system in equilibrium. For this the force polygon must close, and any funicular polygon which corresponds to a pole O must also close.

In particular cases it is sometimes more advantageous to make use of such a funicular polygon in finding any of the applied forces which may be unknown. The points O_1 and O_2 can then be found, and thus the stresses in the rods determined.

138. The force diagram becomes greatly simplified if one of the forces X_2 vanishes. We give the figures below, which the student should think out for himself.

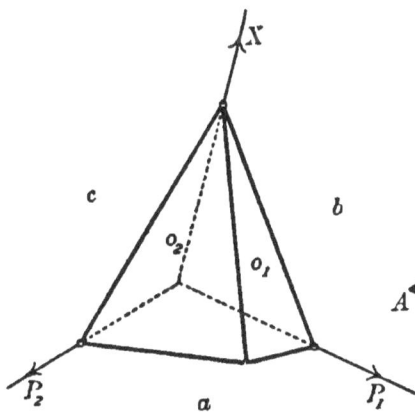 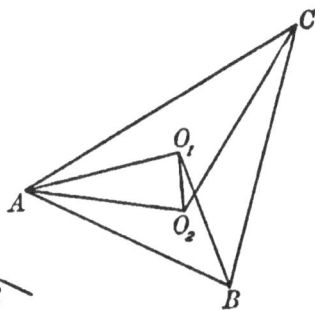

FIG. 149.   FIG. 149 a.

Here it may be advantageous to make use of the fact that the lines of action of P_1, X, P_2 must be concurrent.

If the point of concurrence is inaccessible we might then make use of the funicular polygon.

We give below the figures for the case in which P_1, P_2, and X are parallel.

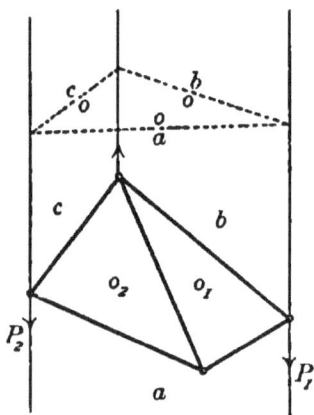

FIG. 150.

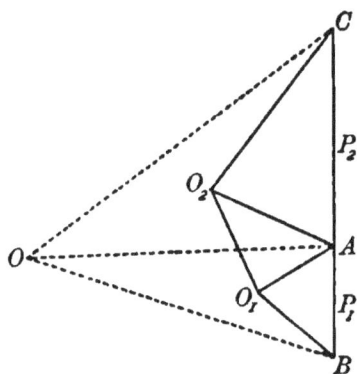

FIG. 150 a.

Or thus:

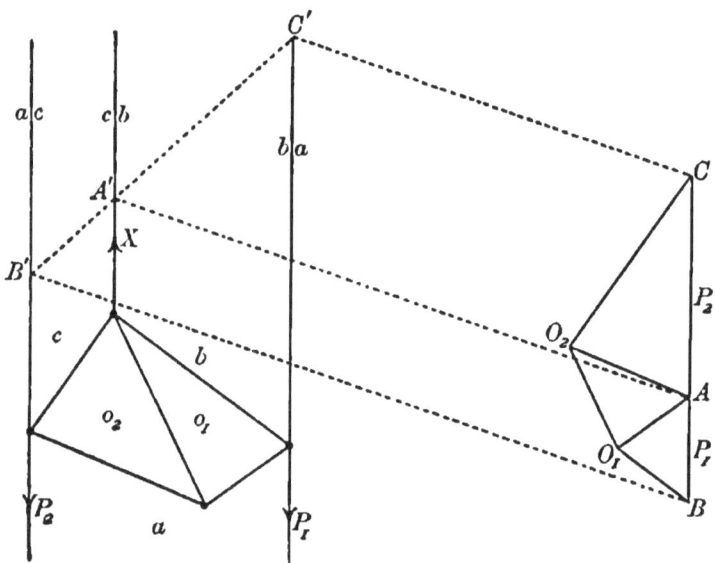

FIG. 151.

The case in which X_1 and X_2 both vanish has been considered in Art. 43.

139. In the general case above, suppose that one side of the framework is fixed in position. The extremities of this rod then become fixed points, and we may suppose the rod itself removed. The figures of Art. 136 then become as below, the rod o_2d having been removed, and the lines of the force diagram which meet in D having all been cut out.

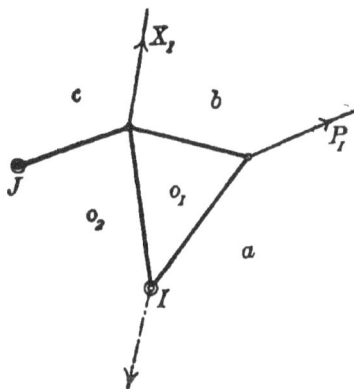
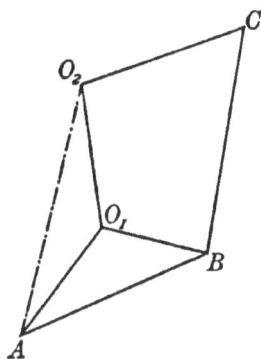

FIG. 152.   FIG. 152a.

The student should think out the problem for himself independently of the general case.

The points I and J are fixed points.

If we join O_2A, we have a straight line representing the reaction at the hinge I.

140. Ex. 1. *Four rods HK, KL, LM, MH, of lengths 3, 4, 3, 4 feet respectively, are freely jointed at their extremities to form a rectangular framework, which is stiffened by another fine light rod connecting the hinges K and M. The framework is capable of turning freely in a vertical plane about the point H which is*

fixed, and rests with KM horizontal under a load of
100 pounds applied at L and an unknown horizontal
force applied at M. Find the magnitude of the force
at M and the action in each rod.

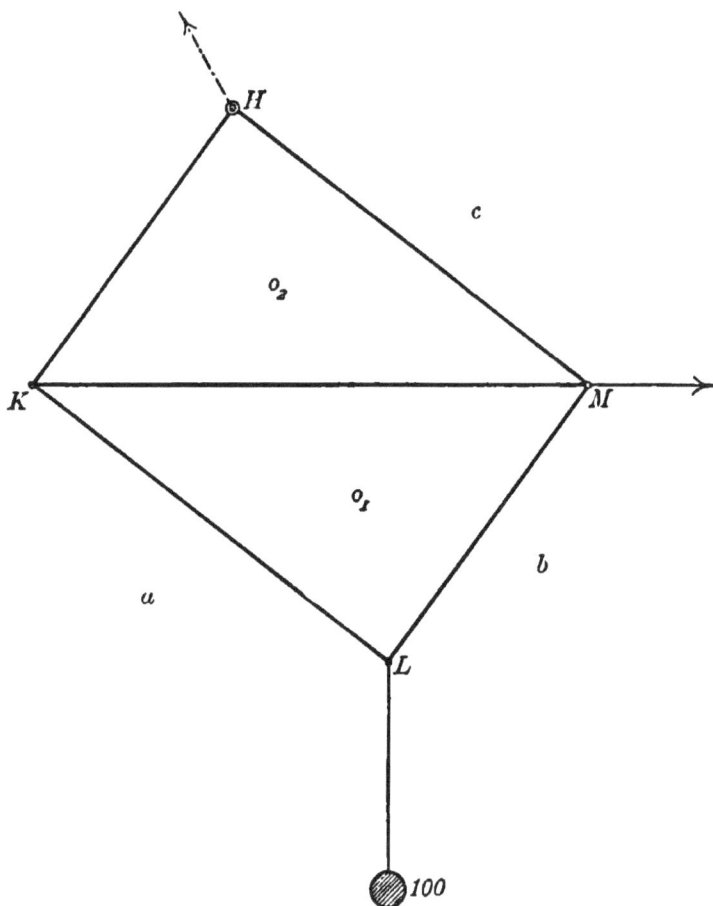

FIG. 153.

Having constructed the space diagram to scale, we
mark it with the letters a, b, c, o_1, o_2, as in the figure.

With any suitable scale, draw AB vertically downwards of length 100 units, to represent the weight of

the mass supported at L. Draw AO_1, BO_1 parallel to
ao_1, bo_1 respectively, thus obtaining the point O_1; draw
AO_2, O_1O_2 parallel to ao_2, o_1o_2 respectively, thus obtaining the point O_2; draw O_2C, BC parallel to o_2c, bc
respectively, thus obtaining the point C.

The triangle of forces for the joint abo_1 is ABO_1A
(this way round); hence the rods KL, LM are tie rods.

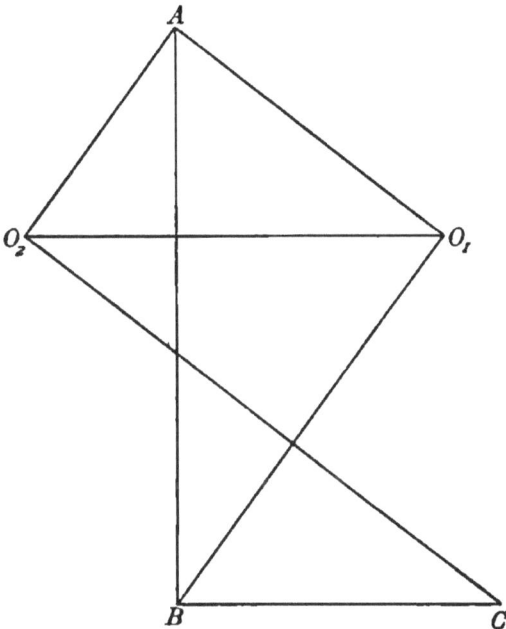

FIG. 153 a.

For the joint o_1bco_2 we have $O_1BCO_2O_1$ (this way round);
hence MH is a tie rod and KM a strut. For the joint
o_1o_2a we have $O_1O_2AO_1$ (this way round); hence KH
is a tie rod.

On measuring the lines of the force diagram, we see that
the force applied at M is 58·3 pounds' weight,
the actions in the rods HK, KL, LM, MH are ties
of 45, 60, 80, 106·7 pounds' weight respectively,

and the action in the rod KM is a strut of 75 pounds' weight.

The force of constraint at H is represented by CA, and it will be found that this line is parallel to the line joining H to the point where the vertical through L meets KM.

141. Ex. 2. *HKLM is a framework of four light rods of given lengths loosely jointed at their extremities. The hinges H and L are connected by means of a fine string of given length. Given loads are attached at K and M and the whole is suspended from H. It is required to find the position of equilibrium, the tension of the string, and the stresses in the rods.*

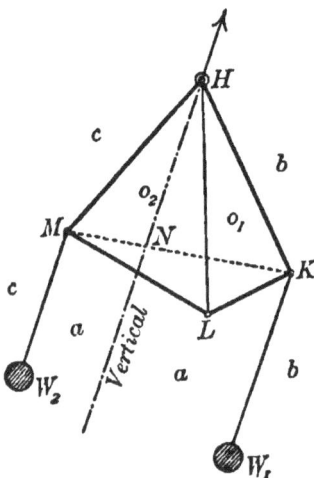

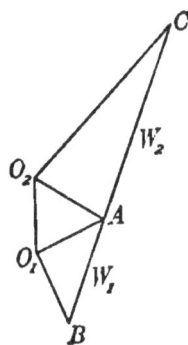

FIG. 154.   FIG. 154 a.

The data are sufficient to enable us to construct the shape of the framework, but not its position relatively to the vertical.

Let W_1 and W_2 be the weights of the loads suspended from K and M respectively. Divide KM in the point N

so that $MN:NK = W_1:W_2$. Then the point N must be vertically below H. This determines the position of equilibrium.

Having lettered the portions of the space diagram as indicated, we draw CAB parallel to HN, making CA, AB of lengths W_2 and W_1 units respectively. Draw CO_2, AO_2 parallel to the rods co_2, ao_2 respectively; this gives us the point O_2. Draw AO_1, BO_1 parallel to the rods ao_1, bo_1 respectively, and we have the point O_1. Then O_2O_1 must be parallel to the string o_2o_1.

On measuring the lines of the force diagram we have all the stresses required.

142. Ex. 3. *The accompanying figure represents a framework of light rods resting on two smooth hori-*

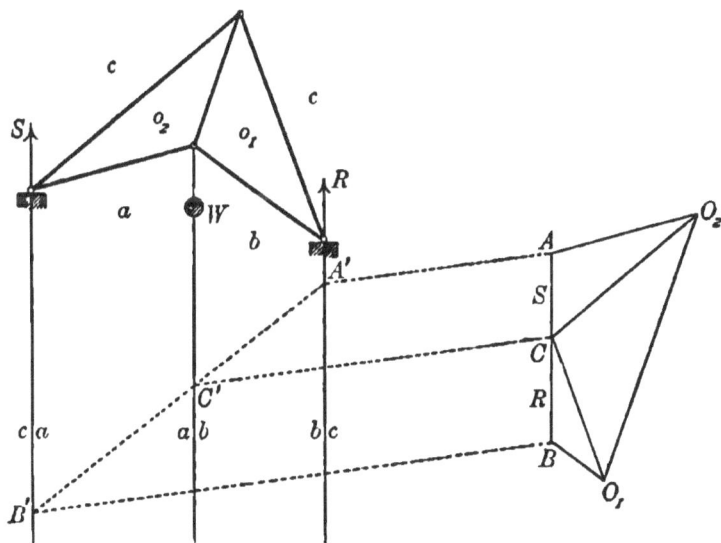

FIG. 155.

zontal supports and loaded at the joint abo_1o_2 with a mass of given weight W. It is required to find the pressures on the supports and the stresses in the rods.

Let R and S be the reactions of the supports acting along the lines bc, ca respectively, as indicated.

Draw AB to represent W, and let two parallel lines through A and B meet bc and ca in A' and B' respectively. Let $A'B'$ meet the line ab in C'. Draw $C'C$ parallel to $A'A$ to meet AB in C. Then BC and CA represent R and S respectively.

We can now construct the triangles of forces BCO_1B, CAO_2C for the joints bco_1, cao_2 respectively. O_1O_2 must then be parallel to the rod o_1o_2, and represent the stress in that rod.

The two lower rods and the middle rod are seen to be *ties*, and the two upper rods *struts*.

143. Ex. 4. *The framework of light rods represented below rests · upon two smooth supports situated in a horizontal line, as indicated in the diagram. Loads*

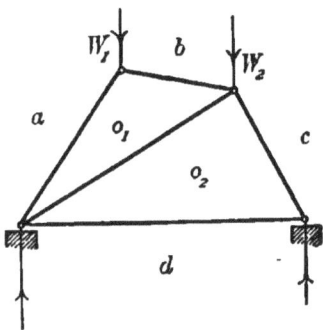

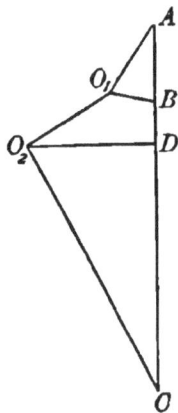

FIG. 156.   FIG. 156 a.

of given weight W_1 and W_2 are applied at the joints abo_1, bco_2o_1, as indicated. It is required to find the pressures on the supports and the stresses in the rods.

Taking AB vertically downwards of length W_1 units, we can construct the triangle ABO_1A, the force diagram for the joint abo_1. Then drawing BC vertically downwards of length W_2 units, we can draw the lines CO_2, O_1O_2, and thus complete the force diagram for the joint bco_2o_1. Then, drawing O_2D parallel to the rod o_2d to meet AC in D, the force diagram is complete.

EXAMPLES XVI.

1. Four rods, of no appreciable weight, and each of length 4 feet, are hinged together at their extremities to form a rhombus $ABCD$, and the hinges A and C are connected by a fine light string of length 7 feet. If the rhombus is suspended from A, and masses, each weighing 1 cwt., are suspended from B and D, determine the tension of the string and the stresses in the rods.

2. $ABCD$ is a framework of four light rods freely jointed together at their extremities, AB and AD being each of length 4 feet, BC and CD each of length 2 feet. The hinge C is connected with A by means of a fine light string of length 5 feet, masses of 100 pounds each are attached at B and D, and the whole is suspended from A. Find the tension of the string and the stresses in the rods.

3. Three fine light rods BC, CD, DB, of lengths 8, 17, 15 inches respectively, are freely jointed at their extremities to form a triangular framework BCD, which is capable of turning freely about the fixed point D. Another light rod AB, of length 9 inches, connects B with a fixed point A, situated 12 inches vertically above D. To the joint C is attached a mass of 300 pounds; find the stresses in the rods.

4. Four light rods, each of length 20 inches, are freely jointed together at their extremities to form a rhombus $ABCD$. The hinges A and C are connected by a fine string, of length 24 inches. Loads of 7 and 25 pounds are applied at B and D respectively, and the whole is suspended from A. Find the position of equilibrium, the tension of the string, and the stresses in the rods.

5. $ABCD$ is a framework of four light rods freely jointed together at their extremities, AB and BC being each of length 20 inches, CD and DA each of length 15 inches. A fine string, of length 2 feet, connects the hinges A and C. A mass of 108 pounds is suspended from B, and the whole is supported at A. What load must be applied at D in order that AC may be vertical? Find also the tension of the string and the stresses in the rods.

6. The framework of Art. 140 is capable of turning freely about H, which is fixed, and rests with MK vertically down-wards, under a load of 100 pounds applied at K and a vertical force at L. Find the magnitude of the force at L and the stresses in the different rods.

7. Four light rods AB, BC, CD, DA, of lengths 10, 17, 17, 10 inches respectively, are freely jointed at their extremities to form the quadrilateral framework $ABCD$, which is stiffened by another light rod, of length 16 inches, connecting the hinges B and D. The framework is capable of turning freely in a vertical plane about a fixed point A, and rests with DB vertically downwards, under a load of 1 cwt. applied at B and a vertical force applied at C. Find the magnitude of the force at C and the stresses in the rods.

8. Four light rods, each of length 5 feet, are freely jointed at their extremities to form a rhombus $ABCD$, which is stiffened by another light rod, of length 6 feet, connecting the hinges B and D. The framework rests with DB vertically downwards, a mass of 192 pounds being suspended from B, and is supported by two forces applied at A and C, in directions perpendicular to DA, DC respectively. Find the magnitudes of these forces and the stresses in the rods.

9. AB, BC, CD are three equal rods, of no appreciable weight, smoothly hinged together at B and C and to fixed points at A and D, the figure forming one half of a regular hexagon, with BC horizontal and below AD. The framework is stiffened by another light rod AC, and loads of 10 and 30 pounds respectively are applied at B and C. Find the stresses in the rods.

10. Four light rods AB, BC, CD, DA are freely jointed together at their extremities to form half of a regular hexagon, the rod DA being double the length of each of the others. The framework is stiffened by another light rod AC, and the whole rests with AD horizontal upon two supports at A and D. Loads of 10 and 30 pounds respectively are applied at B and C. Determine the stresses in the rods and the reactions of the supports.

11. Find the stresses in the rods in the case indicated in Fig. 157, the lengths of the rods being indicated in feet, and the 5-foot rod being vertical.

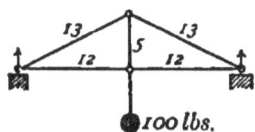

FIG. 157.

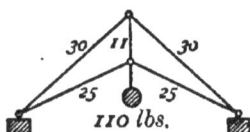

FIG. 158.

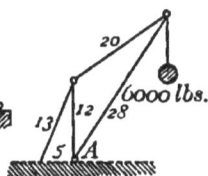

FIG. 159.

12. Find the stresses in the rods in the case indicated in Fig. 158, the 11-foot rod being vertical.

13. Find the stresses in the rods in the case of the derrick crane indicated in Fig. 159, the 12-foot rod being vertical and the ground horizontal. Find also the strain on the hinge A.

14. In the general case considered in Art. 136, suppose that the external forces applied to the framework are all given in position, and that the magnitude of one is known, and of the others unknown. Show how to determine the magnitudes of the other three forces and the stresses in the rods.

- 15. Suppose that P_1 and P_2 are given in magnitude and direction, X_1 in direction only, and X_2 wholly unknown. Show how to find the unknown forces and the stresses in the rods.

16. In the case where X_2 is zero, suppose that P_1 is given completely, P_2 only in direction, and X_1 wholly unknown. Show how to find the unknown forces and the stresses in the rods.

17. Four light rods are freely jointed at their extremities to form a parallelogram $ABCD$, which is stiffened by another light rod connecting the hinges B and D. The framework is capable

D.S.   R

of turning freely in a vertical plane about a fixed point A, and rests with DB vertically downwards, a mass of weight W being suspended from B, and a vertical force applied at C. Show that the force applied at C, the tension of the rod BD, and the force of constraint at A are each equal to $\frac{1}{2}W$, and that the tensions of the rods AB, BC, BD and the compressions of the rods CD, DA are proportional to AB, BC, BD, CD, DA respectively.

18. Four fine light rods are freely jointed at their extremities to form a parallelogram $ABCD$, which is stiffened by another fine light rod connecting B and D. The framework rests with B vertically below E, the middle point of DC, and is loaded at B, being supported by vertical forces at A and C. Prove that the weight of the load at B, the forces at A and C, the tensions of the rods AB, BD, BC, and the compressions of the rods CD, DA are proportional to $3.BE$, BE, $2.BE$, $\frac{1}{2}.AB$, BD, $2.BC$, CD, DA respectively.

19. In the figures of Art. 141, prove that MK is parallel to the straight line joining A to the point of intersection of BO_1 and CO_2.

20. In Art. 141, if HL meets MK in I, and $W_1 : W_2 = MI : IK$, prove that, in the position of equilibrium, HL is vertical, and that, if T is the measure of the tension of the string, $T : W_1 + W_2 = LI : LH$; if, in addition, HM is parallel to KL, then $T = W_2$.

21. Four fine light rods are freely jointed at their extremities to form a convex quadrilateral framework $ABCD$, which is stiffened by another fine light rod connecting B and D. A mass of weight W is suspended from B, and the whole system is supported, with DB vertically downwards, by vertical forces applied at A and C. Prove that, if T is the tension of the rod DB, and E the point where the straight line AC intersects BD, then $T : W = DE : DB$; in particular, if AC bisects DB, then $T = \frac{1}{2}W$.

22. If, in the preceding example, AB is parallel to DC, the tension of the string is equal to the force applied at A.

23. In the figures of Art. 136, prove that the line joining the intersection of AO_1 and DO_2 with the intersection of BO_1 and CO_2 is parallel to the line joining abo_1 to cdo_2.

24. In the figures of Art. 138, prove that the line joining A to the intersection of BO_1 and CO_2 is parallel to the line joining abo_1 to cao_2.

25. Four fine light rods are freely jointed at their extremities to form a convex quadrilateral framework $ABCD$, which is stiffened by another fine light rod connecting B and D. The straight line AC meets BD in E, and the shape of the framework is such that E is the middle point of AC. A mass of weight W is suspended from B, and the whole system is supported with DB vertically downwards, by means of two forces applied at A and C, in directions CD and AD respectively. If T is the tension of the rod DB, prove that $W - T : W + T = DE : EB$.

CHAPTER XVII.

STIFF FRAMEWORKS OF FINE LIGHT RODS SMOOTHLY JOINTED AT THEIR EXTREMITIES.

144. IN the preceding chapter we have confined our attention to the consideration of the equilibrium of a stiff quadrilateral framework, and to the determination of the stresses induced in the different rods. The same method can, however, be applied to all frameworks of fine light rods jointed at their extremities.

In the case of *indeformable* or *stiff* frameworks, *i.e.*, frameworks whose angles do not admit of variation, it is necessary and sufficient for equilibrium that the external forces acting on the framework should form a system in equilibrium. The stresses in the rods are found by considering the equilibrium of each joint separately; if we know the stresses in all but two of the rods which meet at any joint we can determine the stresses in the remaining two by making the polygon of forces for that joint close.

The method will be made clear by working out a few typical examples.

145. Ex. 1. *The framework represented below, resembling a bent crane, is loaded at the joint abo_1 with a mass of given weight W. It is required to determine the stresses in the various rods.*

Draw AB vertically downwards and of length W units. Then AO_1, BO_1 can be drawn parallel to the rods ao_1, bo_1 respectively, thus forming the force triangle ABO_1A (this way round) for the joint abo_1. The rod ao_1 is therefore a *strut* and the rod bo_1 a *tie*.

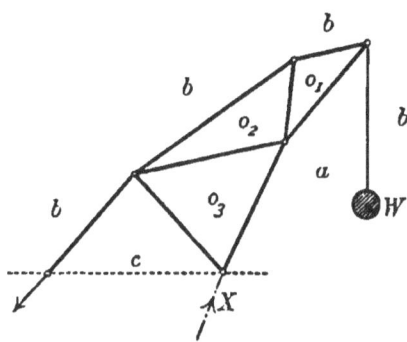

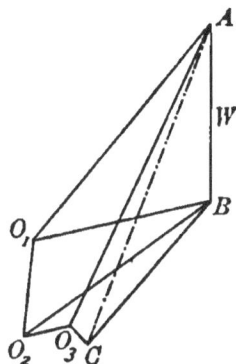

FIG. 160.   FIG. 160 a.

Draw BO_2, O_1O_2 parallel to the rods bo_2, o_1o_2 respectively, and we have the force triangle $O_1BO_2O_1$ (this way round) for the joint o_1bo_2. Thus the rod bo_2 is a *tie* and o_2o_1 is a *strut*.

Now consider the joint $ao_1o_2o_3$. AO_1 and O_1O_2 represent two of the forces acting on this joint. Hence, drawing O_2O_3, AO_3 parallel to the rods o_2o_3, ao_3 respectively, we have the force polygon $AO_1O_2O_3A$ (this way round) for the joint $ao_1o_2o_3$. Thus o_2o_3 and o_3a are both *struts*.

Now consider the joint o_3o_2bc. O_3O_2 and O_2B represent two of the forces acting on this joint. Hence, drawing O_3C, BC parallel to the rods o_3c, bc respectively, we have the force polygon $O_3O_2BCO_3$ (this way round) for the joint o_3o_2bc. Thus the rod bc is a *tie* and co_3 is a *strut*.

$ABCA$ (this way round) is the triangle of forces for the whole framework considered as one rigid body. The action X at the joint ao_3c is represented by CA. The force triangle for this joint is CAO_3C (this way round).

146. *Ex. 2. The framework indicated below is supported at the joints abo_1, bco_3, and supports a mass of given weight W at the joint $cao_1o_2o_3$. It is required to determine the pressures on the supports and the stresses in the rods.*

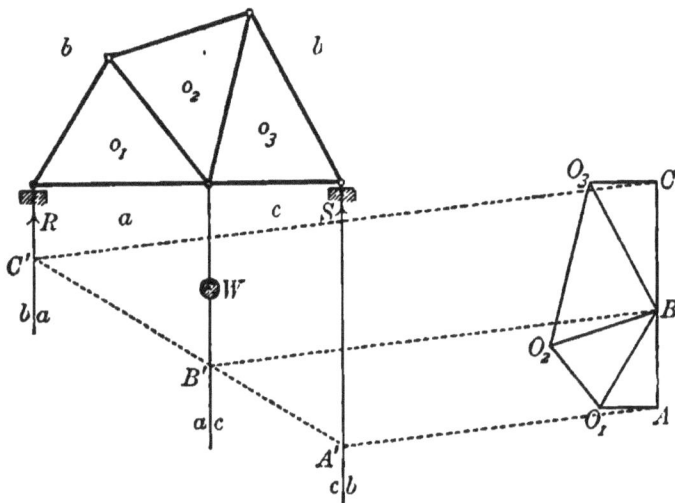

FIG. 161.

Here the only force known is W, but we cannot construct the force polygon for the joint $aco_3o_2o_1$, as the stresses in the four rods which meet at this joint are all unknown. We therefore determine first the pressures R and S between the framework and the supports, by considering the equilibrium of the whole framework as a rigid body.

Draw CA vertically downwards of length W units, and let two parallels through C and A meet the lines ab, bc in C' and A' respectively. Let $C'A'$ meet the line ca in B', and draw $B'B$ parallel to $A'A$ to meet AC in B. Then AB, BC represent the forces R and S which act in the lines ab, bc respectively.

We can now draw the force triangles ABO_1A, BCO_3B for the joints abo_1, bco_3 respectively. Then O_1O_2, O_3O_2 can be drawn parallel to the rods o_1o_2, o_3o_2 respectively, and BO_2 must be parallel to the rod bo_2.

On examining the force diagram for each joint separately, it will be seen that the three top rods are *struts* and the other four *ties*.

147. *Ex. 3. The framework represented below, resembling a portion of a Warren girder, is supported at the joints abo_1, bco_5, and carries loads of given*

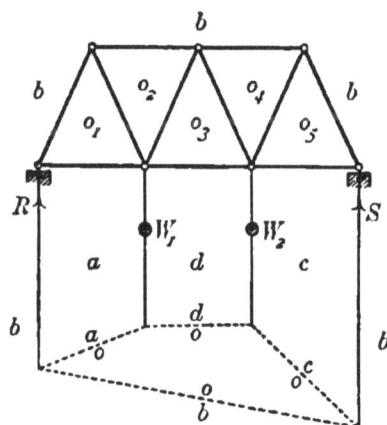

FIG. 162.

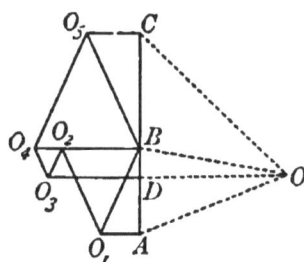

FIG. 162 a.

weight W_1 and W_2 applied at the joints $o_1o_2o_3da$, $o_3o_4o_5cd$ respectively. It is required to determine the pressures on the supports and the stresses in the rods.

We first consider the equilibrium of the whole framework as a rigid body, and thus determine the reactions R and S of the supports. Draw CDA vertically downwards, making CD of length W_2 units, and DA of length W_1 units. Take any pole O, and from any point in the line ab draw the string ao parallel to AO. From the point of intersection of ao and ad draw do parallel to DO. From the point of intersection of do and dc draw co parallel to CO. The straight line ob, drawn from the point of intersection of oa and ab to that of oc and bc, completes the funicular polygon. Draw OB parallel to ob, to meet CA in B. Then $ABCDA$ is the force polygon for the whole system, AB and BC representing R and S respectively.

In constructing the diagram for the stresses in the rods, we commence with the joint abo_1. This gives the triangle ABO_1A (this way round). Then we construct the triangle $O_1BO_2O_1$ (this way round) for the joint o_1bo_2; then the polygon $DAO_1O_2O_3D$ (this way round) for the joint $dao_1o_2o_3$; then the polygon $O_3O_2BO_4O_3$ (this way round) for the joint $o_3o_2bo_4$; then $CDO_3O_4O_5C$ (this way round) for the joint $cdo_3o_4o_5$. Finally, joining BO_5, the triangle $O_5O_4BO_5$ (this way round) is the force triangle for the joint o_5o_4b, and the triangle O_5BCO_5 (this way round) is the force triangle for the joint o_5bc.

148. Sometimes it is not required to determine the stresses in *all* the rods. The following method is then very useful:

Suppose that, in the preceding example, it is required to determine the stress in the rod o_3d. Construct the force polygon $ABCDA$ as before for the equilibrium of the whole system, thus determining R and S.

Draw a line XY intersecting the rods o_4b, o_4o_3, o_3d; and consider the equilibrium of the portion of the framework to the right of this line as one rigid body. The external forces acting upon it are the known forces S and W_2, together with the unknown forces acting at the points U, V, Z. The lines of action of these three unknown forces are all known, and therefore the problem comes under the head of that considered in Art. 118.

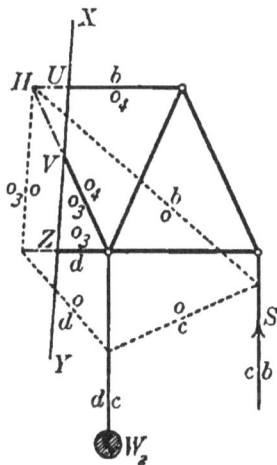

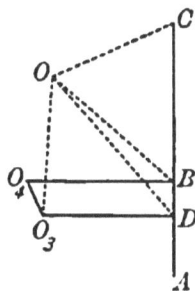

FIG. 163.   FIG. 163 a.

Take a pole O, and draw OB, OC, OD. From H, the point of intersection of the rods o_3o_4, o_4b, draw the line ob parallel to OB. From the point of intersection of ob and bc draw oc parallel to OC. From the point of intersection of oc and cd draw od parallel to OD. From H draw the line oo_3 to the point of intersection of od and do_3, thus completing the funicular polygon. Draw OO_3 parallel to oo_3, to meet the horizontal through D in O_3. Draw BO_4, O_3O_4 parallel to the rods bo_4, o_3o_4

respectively, meeting in O_4. Then $BCDO_3O_4B$ (this way round) is the force polygon for the portion of the framework under consideration. Thus the rod do_3 is in a state of tension, and its tension is represented by DO_3. We have also at the same time determined the actions in the other two rods intersected by the straight line XY.

In applying this method care must be taken not to intersect more than *three* rods. If the stress in one of three rods intersected is known, the stresses in the other two can be found without the aid of a funicular polygon.

149. Another point of practical importance is exhibited in the following example:

Ex. 4. *The framework of Ex. 2 rests on supports as before, and is loaded at each of the other three joints in the manner indicated below. It is required to determine the pressures on the supports and the stresses in the rods.*

We first plan out a force diagram for the whole system, thus determining the pressures R and S on the supports. As the loads are all known we take the forces in the order W_1, W_2, W_3, S, R, determining the unknown forces S and R with the aid of a funicular polygon. The order W_1, W_2, W_3, S, R is not, however, convenient for determining the internal stresses. After determining S and R, we therefore plan out another force diagram, taking the forces in the order R, W_1, W_3, S, W_2, taken one way round the framework, and commencing at a joint where there are only two rods. We can then complete the stress diagram.

For the purpose of the first force diagram, we mark

the lines of action of the forces W_1, W_2, W_3, S, R with the letters fg, gh, hk, kl, lf, as in the figure.

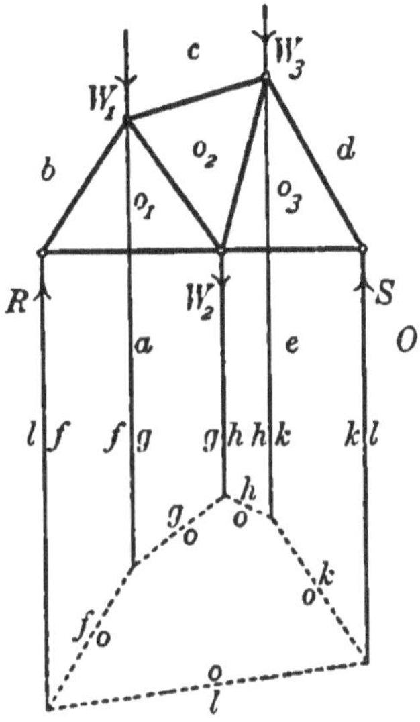

FIG. 164. 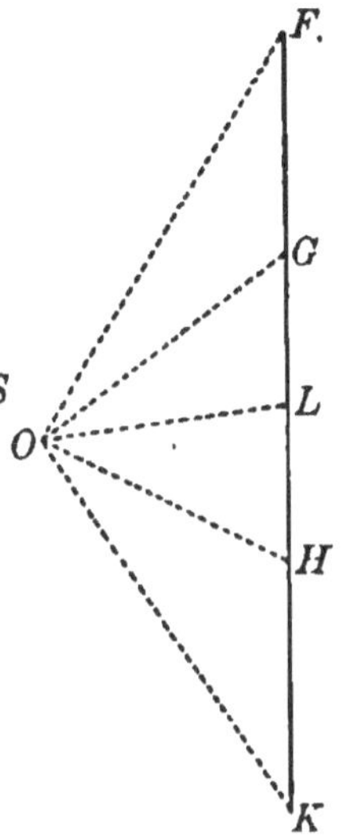   FIG. 164 a. 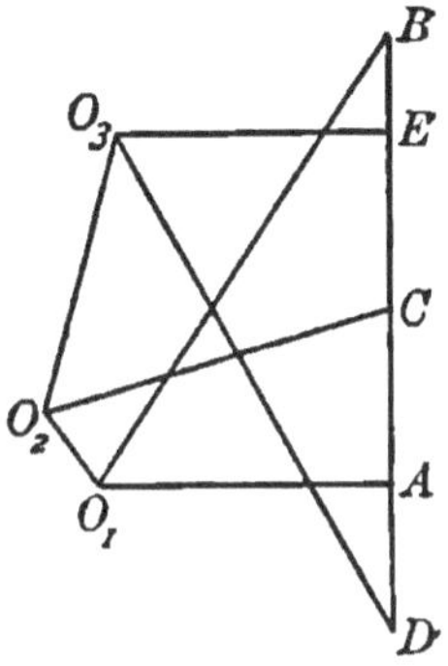   FIG. 164 b.

Take FG, GH, HK to represent W_1, W_2, W_3 respectively, and draw the strings of, og, oh, ok corresponding to the pole O. Draw OL parallel to ol, which completes the funicular polygon, and let OL meet FK in L. Then KL and LF represent S and R respectively.

Now draw another force diagram $ABCDEA$, taking AB, BC, CD, DE equal to and in the same direction as LF, FG, HK, KL, and therefore EA equal to and in the same direction as GH.

Draw BO_1 parallel to the rod bo_1 to meet the horizontal through A in O_1, and DO_3 parallel to the rod do_3 to meet the horizontal through E in O_3. Through O_3 and O_1 draw O_3O_2, O_1O_2 parallel to the rods o_3o_2, o_1o_2 respectively, to meet in O_2. Then CO_2 must be parallel to the rod co_2.

The triangle of forces for the joint abo_1 is ABO_1A (this way round). The force diagram for the joint o_1bco_2 is $O_1BCO_2O_1$ (this way round); for the joint $o_1o_2o_3ea$, $O_1O_2O_3EAO_1$ (this way round); for the joint o_3o_2cd, $O_3O_2CDO_3$ (this way round); for the joint o_3de, O_3DEO_3 (this way round).

EXAMPLES XVII.

1. $OABCD$ is a framework of light rods smoothly jointed at their extremities; the rods OA, OB, OC, OD being each of length 25 inches; the rods AB, CD each of length 14 inches; and the rod BC of length 30 inches. Two masses, weighing 100 pounds each, are suspended from A and D, and the whole is supported at O. Find the stresses in the rods.

2. In Art. 146, each of the triangles o_1, o_2, o_3 is equilateral. Find the stresses in the rods when the load is 1000 pounds.

3. In the same Example, each of the lower rods is of length 6 feet; the top middle rod is also of length 6 feet, and the other four rods are each of length 5 feet. Find the stresses in the rods when the load is 400 pounds.

4. Draw the stress diagram for the framework considered in Art. 147, when three equal loads are applied at the upper joints, and no loads are applied at the lower joints.

5. In the framework considered in Art. 147, each of the horizontal rods is of length 2 feet 6 inches, and each of the others of length 3 feet 3 inches. Find the stresses in the rods o_4o_5, bo_4, co_5, when the only load supported is 1 ton at the joint $o_1o_2o_3da$.

6. Draw the stress diagram for a Warren Girder of 4 *bays* (instead of 3, as in Art. 147), when the only load applied is a mass of weight W at the middle lower joint.

7. In the preceding example the load is 1 ton, each of the horizontal rods is of length 2 feet 6 inches, and each of the other rods of length 3 feet 3 inches; find the stresses in all the horizontal rods, and examine whether the other rods are *struts* or *ties*.

8. Draw the stress diagram for the framework indicated in Fig. 165.

9. If, in Fig. 165, each of the horizontal rods is of length 4 feet, each of the uprights 3 feet, and each of the others 5 feet, find the stresses in the three rods intersected by the line XY, each of the three loads being 1 ton.

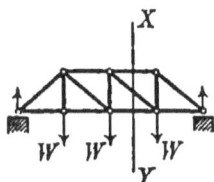

FIG. 165.

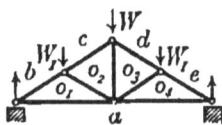

FIG. 166.

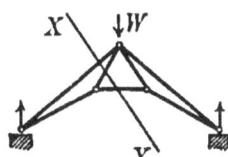

FIG. 167.

10. Draw the stress diagram for the framework indicated in Fig. 166. Notice that the points O_1 and O_4 coincide.

11. Draw the stress diagram for the framework indicated in Fig. 167.

12. Find the stresses in the three members intersected by the line XY in Fig. 167, the rods which meet at the vertex being of lengths 10, 5, 5, 10 feet, the horizontal rod of length 6 feet, the span 16 feet, and the load being 1 cwt.

13. Draw the stress diagram for the framework indicated in Fig. 168. Notice that the points O_1 and O_5 coincide.

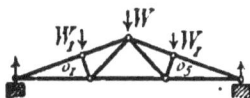

FIG. 168.

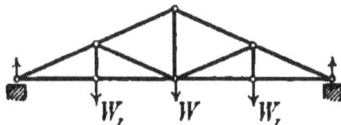

FIG. 169.

14. Draw the stress diagram for the framework indicated in Fig. 169.

270 STIFF FRAMEWORKS.

15. Draw the stress diagram for the framework indicated in Fig. 170, the whole system being suspended from the point H.

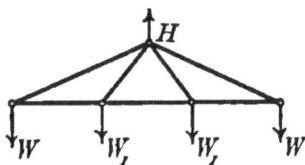

FIG. 170.

16. In the figures of Art. 148, prove that O_4C is parallel to the line joining the point $o_3o_4o_5cd$ to the intersection of bc and bo_4. Hence give a method for determining the stresses in the intersected rods without drawing a funicular polygon.

CHAPTER XVIII.

SYSTEMS OF RODS, SOME, OR ALL, OF WHICH ARE ACTED UPON BY FORCES NOT AT THEIR EXTREMITIES.

150. The accompanying figure represents three rods of a framework smoothly jointed at their extremities. The rods are drawn separated, so that the forces acting on each rod may be clearly indicated. The space on the one side of the rods is named o, and on the other sides the spaces are marked a, b, c, so that the three rods are denoted by oa, ob, oc. Let the hinges be denoted by the numbers 1, 2, 3, 4, as indicated.

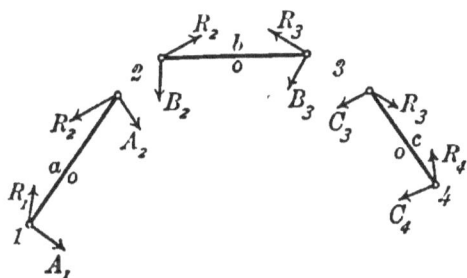
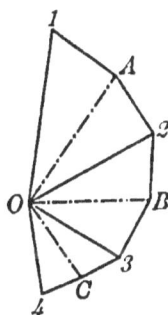

FIG. 171.   FIG. 171 a.

Suppose that each rod is acted upon by a system of forces in addition to the constraints at the hinges.

Let the forces acting on the rod oa (apart from the constraints at its extremities) be resolved into two forces A_1 and A_2 acting at its extremities 1 and 2 respectively. Similarly, let the forces on the rods ob and oc be resolved into the pairs of forces B_2, B_3 and C_3, C_4, as indicated.

Let R_1 be the action at the joint 1 upon the rod oa, R_2 the action at the joint 2 upon the same rod. Then R_2 in the opposite direction is the action at the joint 2 upon the rod ob. Similarly, let R_3 and R_4 be the actions at the joints 3 and 4 respectively.

Consider the equilibrium of the rod oa. The resultant of R_1 and A_1 must balance the resultant of R_2 and A_2. Therefore ao must be the line of action of these two resultants. If then we draw $O1$, $1A$ to represent R_1 and A_1 respectively, OA will have to be parallel to oa, and AO will represent the resultant of R_2 and A_2. Hence if $A2$ represents A_2, then $2O$ must represent R_2. Thus the force diagram for the rod ao is $O1A2O$ (this way round), in which OA is parallel to the rod ao, and $1A$, $A2$ represent A_1 and A_2 respectively.

Similarly, the force diagram for the rod ob is $O2B3O$, in which OB is parallel to the rod ob, and $2B$, $B3$ represent B_2 and B_3 respectively. Also the force diagram for the rod oc is $O3C4O$, in which OC is parallel to the rod oc, and $3C$, $C4$ represent C_3 and C_4 respectively.

The actions at the hinges 1, 2, 3, 4 are given by $O1$, $O2$, $O3$, $O4$ respectively, the direction being determined according to which rod is under consideration.

The student should notice that the lines 12, 23, 34 of the force diagram represent the resultants of the forces (other than the constraints at the hinges) that act upon the rods *12, 23, 34* respectively.

We may, if we please, resolve the forces upon each rod into parallel forces at its extremities. In this case *1A2, 2B3, 3C4* become straight lines.

The rods may be replaced by rigid bodies of any shape hinged together at the points *1, 2, 3, 4,* and the same piece of work applies, the straight lines *oa, ob, oc* being drawn connecting the hinges.

151. The question may be asked, "What forces do OA, OB, OC represent?" The student may be inclined to say that these lines represent the stresses in the rods *oa, ob, oc* respectively. *This is not so.* If the forces acting on the rod *oa* were actually A_1 and A_2 at its extremities, the rod would be in a state of direct compression or tension, and OA would represent the strain at every point of the rod. But in the actual state of affairs, the strain is different at different points of the rod, and the tendency of the forces is to bend the rod as well as to compress it or stretch it. In replacing the forces acting on the rod by an equivalent system, we do not interfere with the equilibrium of the rod as a rigid body, but we *do* interfere with the nature of the internal stresses induced. We make no endeavour to interpret the meaning of the lines OA, OB, OC; the nature of the internal stresses in such cases is beyond the scope of the present volume.

152. As a particular case of the above, suppose that four uniform heavy rods are freely jointed together, and hang in a vertical plane, the points *1* and *5* being fixed.

Let the weights of the rods *oa, ob, oc, od* be given. Suppose also that the rods *ob, oc* are given in position,

D.S.   S

and that it is required to find the positions of the rods
oa, od and the actions at the hinges.

The weight of each rod can be broken up into half
its weight applied at each end. Hence, in constructing
the force diagram, draw $1A2, 2B3, 3C4, 4D5$, to represent
the weights of the rods oa, ob, oc, od respectively,
A, B, C, D being the middle points of the lines $12, 23$,
$34, 45$ respectively.

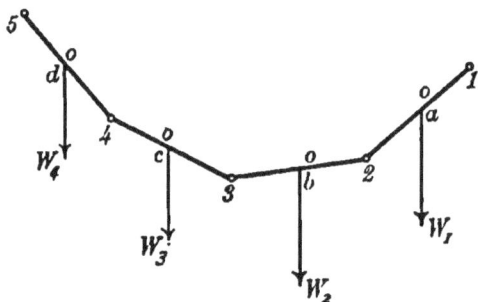
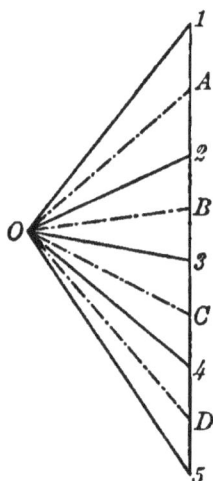

FIG. 172.   FIG. 172 a.

As the rods ob, oc are given in position, we can draw
BO, CO parallel to these rods respectively, and this
determines the point O. Join OA, OD, and we have
the directions of the rods oa, od. Also $O1$, $O2$, $O3$,
$O4$, $O5$ give the actions at the hinges $1, 2, 3, 4, 5$ respectively. The force diagram for the rod oa is $12O1$ (this
way round), so that $2O$ represents the action of the hinge
2 upon the rod oa. Similarly for the other hinges.

153. We have hitherto supposed that all external
forces are applied to the rods themselves, the hinges

being left perfectly free, so that the rods act and react upon one another directly through the hinges. Let us now suppose that a hinge consists of a separate piece (a small pin for instance) of no appreciable size or weight, and that external forces are applied directly upon the hinges. Each hinge is supposed to be perfectly smooth, and its effect upon any rod is to compel the extremity of the rod to remain in a definite position by a direct push or pull exerted upon the rod at its extremity. We shall now have to consider the equilibrium of each hinge as well as of each rod, and the actions of a hinge upon two adjoining rods will not now be equal and opposite.

154. Take the system of rods of Art. 150 acted upon by the same external forces, and in addition let forces F_2 and F_3, in the directions indicated, act upon the

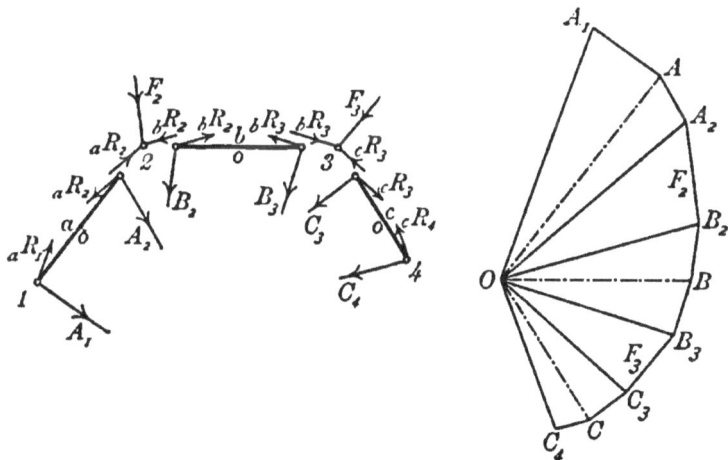

FIG. 173.   FIG. 173 a.

hinges *2* and *3* respectively. In the diagram the rods and hinges are drawn separated, so as to indicate clearly the forces acting upon the separate pieces.

Let the action between the hinge *2* and the rod *oa* be $_aR_2$, and let the action between the hinge *2* and the rod *ob* be $_bR_2$, a similar notation denoting the actions at the other hinges.

In constructing the force diagram, we have for the rod *oa* the figure $A_1AA_2OA_1$ (this way round), in which OA_1 represents $_aR_1$, while A_1A and AA_2 represent A_1 and A_2 respectively, A_2O represents $_aR_2$, and OA is parallel to the rod *oa*. For the hinge *2* we have the triangle OA_2B_2O (this way round), in which A_2B_2 represents F_2, and B_2O represents $_bR_2$. For the rod *ob* we have OB_2BB_3O (this way round), in which B_2B and BB_3 represent B_2 and B_3 respectively, B_3O represents $_bR_3$, and OB is parallel to the rod *ob*. For the hinge *3* we have the triangle OB_3C_3O (this way round), in which B_3C_3 represents F_3, and C_3O represents $_cR_3$. For the rod *oc* we have OC_3CC_4O (this way round), in which C_3C and CC_4 represent C_3 and C_4 respectively, C_4O represents $_cR_4$, and OC is parallel to the rod *oc*.

This piece of work, of course, includes that of Art. 150. If we make F_2 and F_3 each zero, the points A_2 and B_2 will coincide with the point *2* of Art. 150, and the points B_3 and C_3 with the point *3* of that article.

155. It is important that the student should notice that if we suppose the forces A_2, F_2, B_2 collected together and applied at the joint *2*, we get the correct position for the points A and B of the force diagram. Thus, if we do not require to know the exact nature of the constraints in the immediate neighbourhood of the joint, we do not trouble to separate the forces into those which act on the rod *oa*, those which act on the joint, and those which act on the rod *ob*. The

straight line AB, in fact, represents the resultant of
A_2, F_2, B_2.

The student should carefully think over the two
examples which here follow. In the first of these, we
have worked out the problem twice over on two
different suppositions, but they lead to the same diagram.
In the second, we require only the tensions of two
strings, and do not trouble ourselves with the nature
of the actions at the joints.

156. *Ex. 1. Two heavy uniform rods of weight
w and w' are smoothly hinged together at the point 2,
and to fixed points at the points 1 and 3, as represented in the figure on the left. A mass of weight W
is supported at the point 2. It is required to find
the stresses at the hinges.*

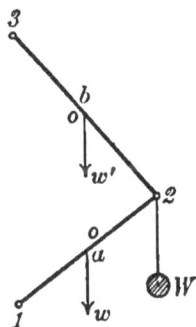
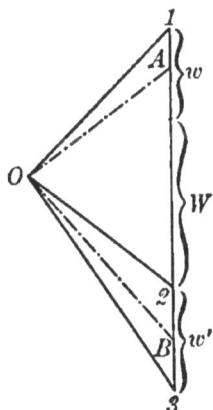
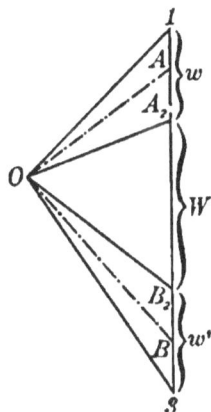

FIG. 174.   FIG. 174 *a*.   FIG. 174 *b*.

First, suppose that the mass of weight W is attached
to the rod oa at a point indefinitely close to the hinge *2*.

The weight of each rod can be broken up into half
its weight at each extremity.

Draw *1A2B3* vertically downwards, taking *1A, A2, 2B, B3* to represent $\frac{1}{2}w$, $\frac{1}{2}w+W$, $\frac{1}{2}w'$, $\frac{1}{2}w'$ respectively. Draw *AO, BO* parallel to the rods *ao, bo* respectively, meeting in *O*. Then *O1, O2, O3* give the actions at the hinges *1, 2, 3* respectively.

Secondly, suppose that the mass of weight *W* is attached directly to the hinge *2*, which is a separate piece.

Draw *1A, AA₂, A₂B₂, B₂B, B3* to represent $\frac{1}{2}w$, $\frac{1}{2}w$, W, $\frac{1}{2}w'$, $\frac{1}{2}w'$ respectively. Draw *AO, BO* as before. Then the actions at the hinges *1* and *3* are given by *O1, O3* respectively, and are the same as in the first case. The action between the hinge *2* and the rod *oa* is given by OA_2, and the action between the same hinge and the rod *ob* is given by OB_2, and is the same as in the first case.

157. Ex. 2. *Three equal uniform rods FG, GH, HK, each of weight w, are freely jointed together at G and H, and laid in a vertical plane upon a smooth hori-*

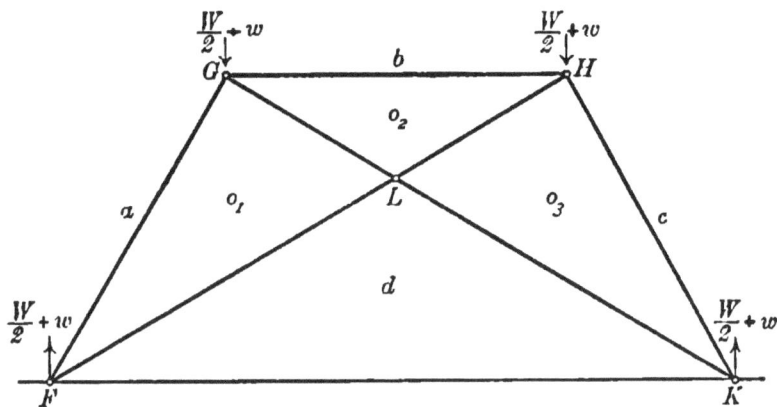

FIG. 175.

zontal table at F and K. Two fine light strings FH, GK keep the system in the form of one-half of a regular hexagon, and a mass of weight W is placed at the middle point of GH. Determine the tensions of the strings.

The reactions at F and K are evidently each equal to $\frac{1}{2}W + \frac{3}{2}w$. The weight of each rod may be replaced by $\frac{1}{2}w$ acting at each of its extremities. Hence, for the purpose of finding the tensions of the strings, we may suppose that the system is a jointed framework of fine light rods, in equilibrium under the influence of forces $\frac{1}{2}W + w$ acting vertically upwards at F, $\frac{1}{2}W + w$ vertically downwards at G, $\frac{1}{2}W + w$ vertically downwards at H, and $\frac{1}{2}W + w$ vertically upwards at K.

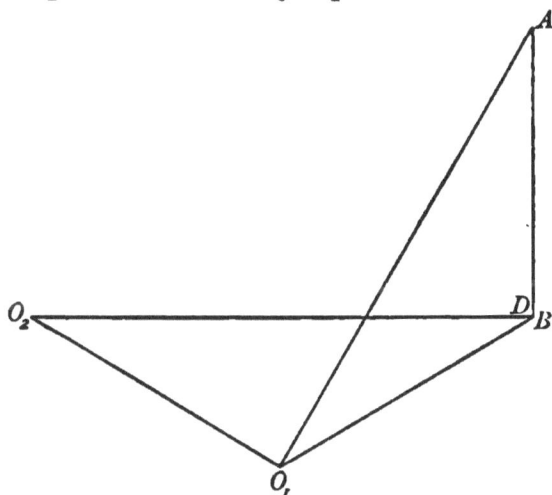

FIG. 175a.

Also, it is convenient for the purpose of constructing a force diagram to suppose the strings knotted together at the point L where they intersect, so that we may look upon the two strings as four separate members attached together at the point L. This will evidently put no additional strain upon either string; the force

diagram for the point L will be a parallelogram, giving
the tension of the two parts of the same string as equal
in magnitude.

Now mark the portions of the space diagram with
the letters a, b, c, d, o_1, o_2, o_3, as indicated. Draw DA
vertically upwards to represent $\frac{1}{2}W+w$, and let AO_1,
DO_1, be drawn parallel to ao_1, do_1 respectively. This
gives the point O_1.

Draw AB vertically downwards to represent $\frac{1}{2}W+w$;
evidently B coincides with D. Draw BO_2, O_1O_2 parallel
to bo_2, o_1o_2 respectively. This gives the point O_2.

There is no necessity to draw any more of the force
diagram. O_1O_2 and DO_1 represent the tensions of the
strings.

It is easy to see that the figure ADO_1O_2 is also
one half of a regular hexagon. Hence the tension of
each string is $\frac{1}{2}W+w$.

158. *Ex. 3. Three uniform rods HK, KL, LM, of
lengths 15, 14, 15 inches respectively, and of weights
W, W', W respectively, are freely hinged together at K
and L, and supported from a fixed point F by means
of three fine light strings FH, FG, FM of lengths 20,
24, 20 inches respectively, the point G being the middle
point of KL. The system rests with FG vertical and
KL horizontal. Determine the tensions of the strings,
and the actions at the hinges.*

Having constructed the space diagram to scale, let
the lines HK, KL, LM, MF, FH be marked oa, ob, oc,
od, oe respectively, and let the hinges K and L be
called *1* and *2* respectively.

Draw a straight line $E1$ vertically downwards to
represent W, and take A the middle point of $E1$. Let

straight lines through E and A, parallel to eo, ao respectively, meet in O. Then $OEA1O$ (this way round) is the force diagram for the rod oa.

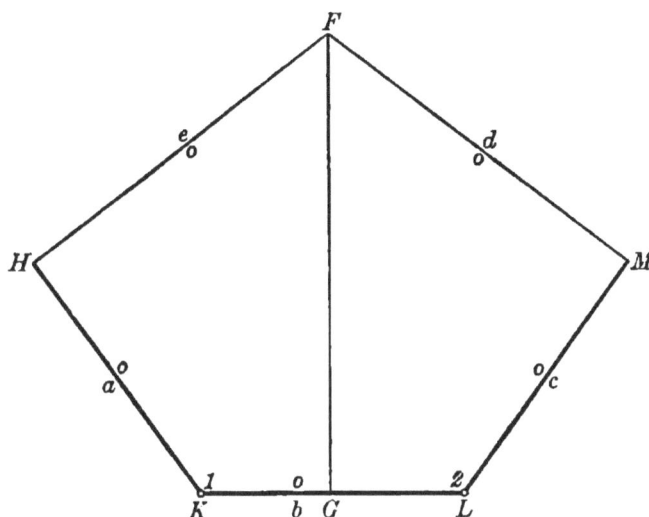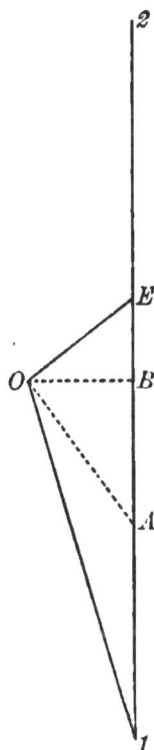

FIG. 176.   FIG. 176 a.

Draw OB parallel to ob to meet $E1$ in B, and produce $1B$ to 2, making $B2 = 1B$. Then $O1B2O$ (this way round) is the force diagram for the rod ob. The straight line $1B2$ represents the resultant of the forces (other than the actions at the joints) which act upon the rod ob. Hence, if T be the tension of the vertical string, $1B2$ represents $T - W'$.

On measurement, we find that the lengths of the

lines OE, $O1$, $1B2$ are respectively $\cdot3\,W$, $\cdot85\,W$, $1\cdot64\,W$; also the angle $O1B = 16\frac{1}{4}°$.

Hence the tensions of the strings FH, FG, FM are respectively $\cdot3\,W$, $1\cdot64\,W + W'$, $\cdot3\,W$; also the actions at the hinges K and L are each $\cdot85\,W$, and are each inclined at an angle of $16\frac{1}{4}°$ to the vertical.

159. Ex. 4. *Two uniform rods KL, LM, each of weight W and length 5 feet, are freely jointed together at L, and connected with a fixed point H by means of two fine light strings KH, HM of lengths 8 and 6 feet respectively. Another fine light string connects H*

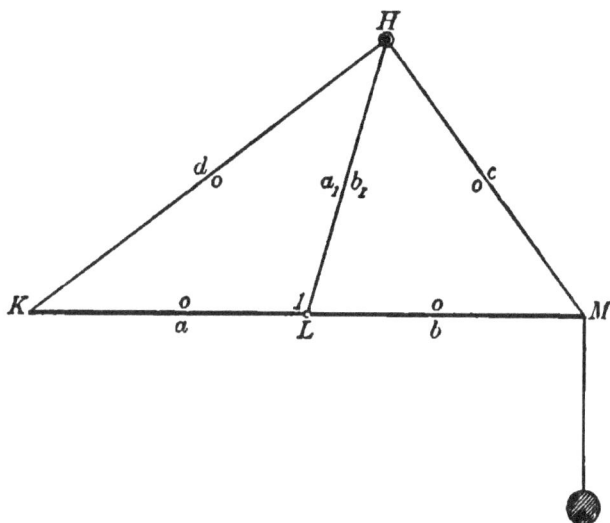

FIG. 177.

with L and is of such a length that when the strings are all tight KL and LM are in one straight line. Find what load must be applied at M in order that, in the position of equilibrium, KLM may be horizontal; determine also the tensions of the strings.

Having constructed the space diagram to scale, mark the lines *KL, LM, MH, HK* with the letters *oa, ob, oc, od* respectively.

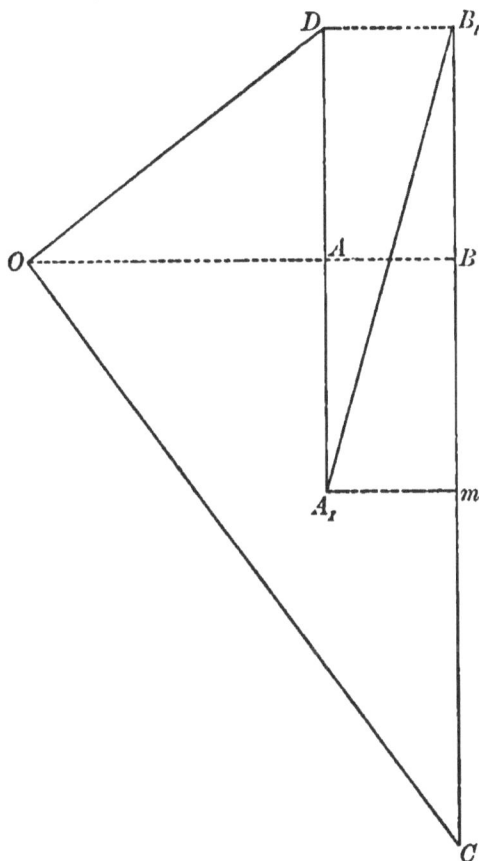

FIG. 177 *a.*

We will suppose that the string *HL* is attached to the hinge at *L*, which is a separate piece. We denote the hinge by the figure *1*, and mark the straight line *HL* with the letters $a_1 b_1$.

With any suitable scale, draw DA_1 vertically down-wards to represent *W*, and bisect in *A* the line so drawn. Let the straight line through D parallel to do meet the horizontal through A in O. Then $ODAA_1O$ (this way round) is the force diagram for the rod oa. Draw A_1B_1 parallel to a_1b_1. The point B_1 is at present unknown; it will have such a position that, if B_1B is drawn vertically downwards to meet OA in B, B_1B must represent $\frac{1}{2}W$. Hence, to obtain the position of B_1, draw DB_1 parallel to OA to meet A_1B_1 in B_1. Then OA_1B_1O (this way round) is the force diagram for the hinge 1.

Draw OC parallel to oc, to meet the vertical through B_1 in C. Then OB_1BCO (this way round) is the force diagram for the rod ob.

The line BC represents the sum of half the weight of the rod ob and the weight of the load applied at M. Hence, drawing A_1m horizontal to meet BC in m, we see that mC represents the weight of the load supported at M.

On measuring the lines mC, OD, A_1B_1, CO we find that

the weight of the load at $M = \cdot78\,W$,
the tension of $HK = \cdot83\,W$,
the tension of $HL = 1\cdot04\,W$,
the tension of $HM = 1\cdot60\,W$.

160. Ex. 5. *Two uniform beams HL, KL, of lengths 7 ft. 6 ins. and 6 feet 6 ins. respectively, and weighing 16 and 12 pounds respectively, are freely jointed at L, and rest in a vertical plane upon a smooth horizontal plane at H and K. A fine light cord MN, of length 4 feet 8 ins., is attached at its extremities to the two beams at the points M and N which divide LH and LK respectively each in the ratio 2 : 1. Find the tension of* *the string, the reactions at H and K, and the action at the hinge.*

Having constructed the space diagram we proceed to find first of all the reactions at H and K. For this purpose we consider the equilibrium of the whole system as one rigid body. Let the lines HL, LK and the verticals through K and H be marked oa, ob, oc, od respectively, and let the hinge L be denoted by 1. Let the verticals through the middle points of HL and LK be marked de, ec respectively.

Draw DE, EC vertically downwards and of lengths 16 and 12 units respectively. Take away pole O' and construct the sides do', eo', co' of a funicular polygon corresponding to the pole O'. Draw the string oo', which completes the funicular polygon, and the line $O'O$ parallel to oo' to meet DC in O. Then CO, OD represent the reactions at K and H respectively.

Now consider the equilibrium of the rod oa alone. We may take the weight of the rod as equivalent to 8 pounds' weight acting at H and 8 pounds' weight acting at K; also the tension T of the string is equivalent to $\frac{2}{3}T$ at H and $\frac{1}{3}T$ at L. Hence, bisecting DE in m, draw OA parallel to oa to meet in A the horizontal through m. Then mA represents $\frac{2}{3}T$. Produce mA to n, making $An = \frac{1}{2} . mA$, and draw $n1$ vertically downwards of length 8 units. Then $ODmAn1O$ (this way round) is the force diagram for the rod oa.

We have completed the figure so as to show the force diagram for the rod ob, but this is unnecessary.

On measurement, we find that CO, OD, mn, $O1$ are of lengths 15, 13, 5·6, 6·4 respectively; also the angle $O1n$ is 62°.

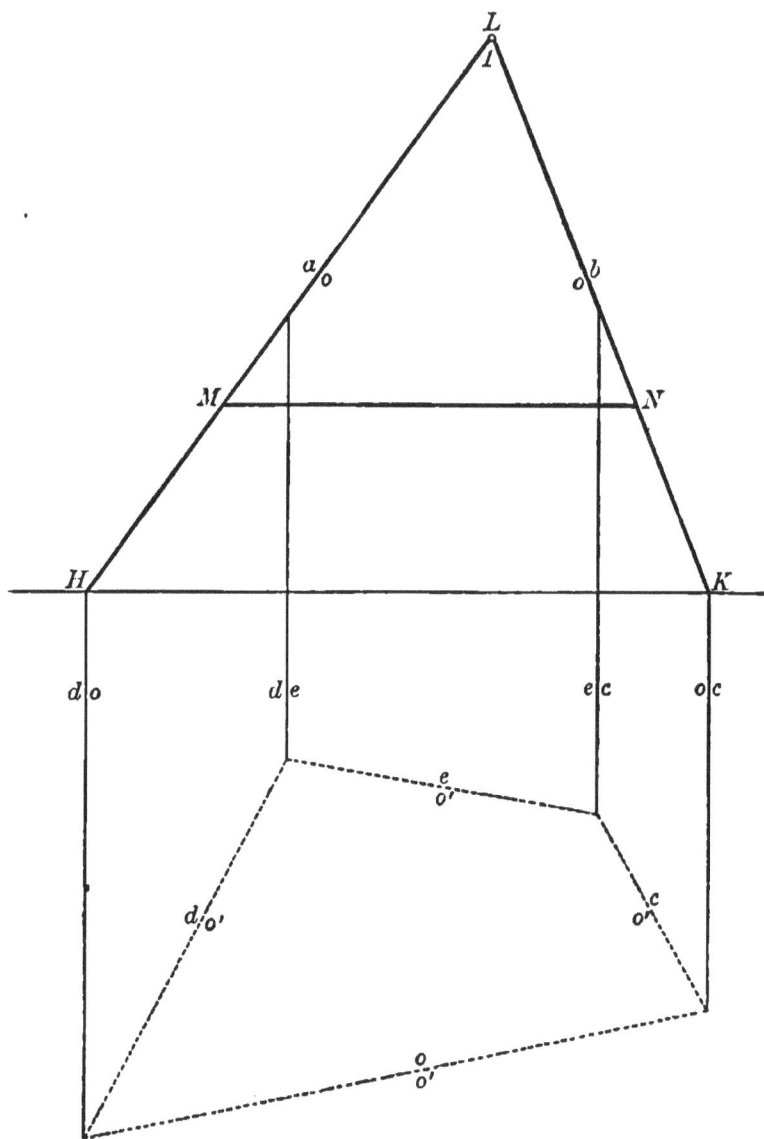

FIG. 178.

Hence the tension of the string and the actions at H and K are respectively 5·6, 13, 15 pounds' weight; also

the action at the hinge L is 6·4 pounds' weight in a
direction making an angle of 62° with the vertical.

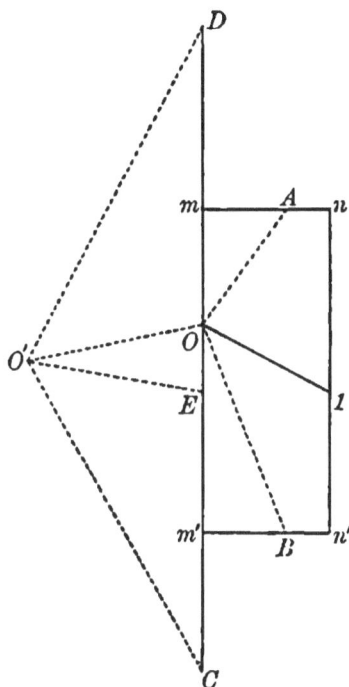

FIG. 178 a.

Another method of solving this example is considered
in the next chapter.

161. Ex. 6. *Four uniform heavy rods are freely
jointed at their extremities to form the quadrilateral
framework indicated below. A fine string, connecting
two opposite hinges, keeps the quadrilateral in a given
shape, and the whole rests in a vertical plane with the
rod oc held in a given position. It is required to
determine the tension of the string.*

Draw $A_1 2$, $2B_3$ to represent the weights of the rods
oa, ob respectively, and take A the middle point of $A_1 2$

and B the middle point of $2B_3$. Draw AO, BO parallel to the rods oa, ob respectively, to meet in O.

The weight of the rod od will be represented by some straight line $4DD_1$, drawn so that D is the middle point of $4D_1$ and OD is parallel to od, and also D_1A_1 must be parallel to the string, and must represent its tension. Hence, to complete the force diagram, draw $A_1D'4'$ vertically upwards, so that $4'A_1$ represents the weight

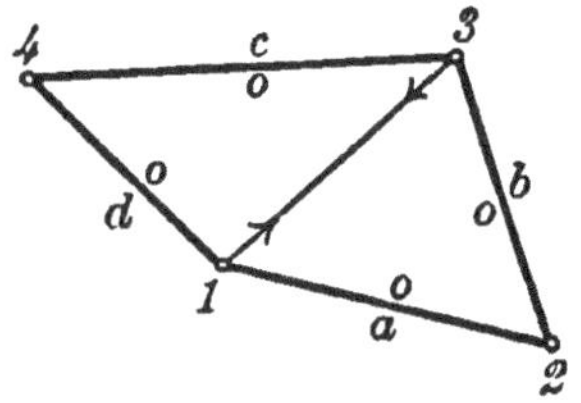

FIG. 179. 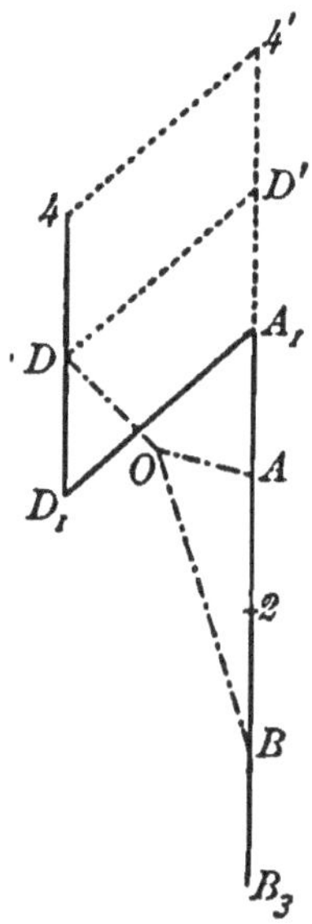   FIG. 179 a.

of the rod od and D' is the middle point of $4'A_1$. Draw $D'D$ parallel to the string to meet in D the line drawn through O parallel to od, and let the straight lines through A_1 and $4'$ parallel to the string, meet the vertical through D in D_1 and 4 respectively. Then, measuring D_1A_1, we have the tension of the string. The rest of the force diagram can be easily completed, and the actions at the hinges determined.

162. **Ex. 7.** *The framework indicated below is composed of fine light rods, freely jointed at their extremities, and is at rest under the influence of forces P, Q, R, applied as indicated. It is required to determine the stresses in the rods, the hinges a and γ being fixed.*

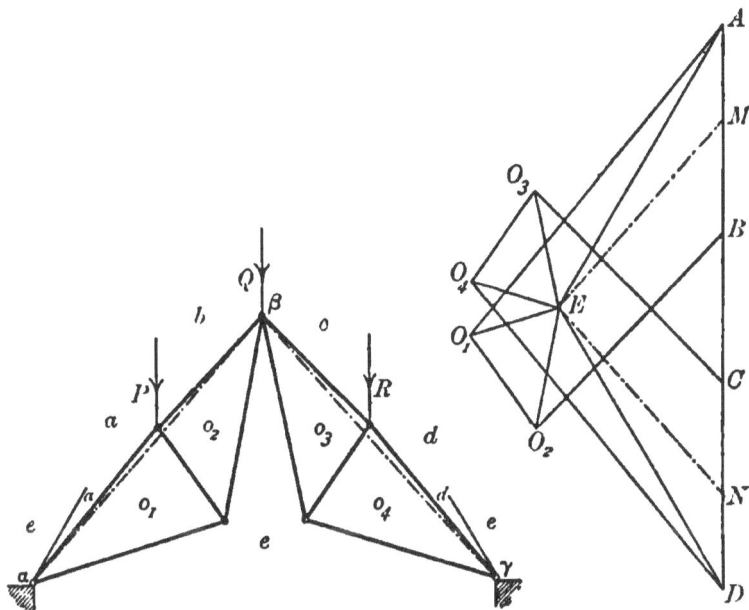

FIG. 180.   FIG. 180 a.

This is a framework like those treated of in the preceding chapter, but in constructing the force diagram we here meet with a difficulty which can be overcome by the methods of the present chapter. At every joint there are more than two unknown forces, so that we cannot at the outset construct the force polygon for any joint.

The reactions at the hinges a and γ are in some unknown directions which will be denoted by *ae*, *de* respectively. We can find these reactions in the following manner:

Consider the triangles o_1 and o_2 as one single rigid body, and the triangles o_3 and o_4 as another rigid body. Draw the straight lines joining the hinges a, β and β, γ. Draw AB, BC, CD to represent the forces P, Q, R applied in the lines ab, bc, cd respectively.

Divide AB in M, so that AM, MB represent components of P through a and β respectively; and divide CD in N, so that CN, ND represent components of R through β and γ respectively. These can be done by Art. 72.

Draw ME, NE parallel to $a\beta$, $\beta\gamma$ respectively, to meet in E. Then, by the methods laid down in this chapter, EA and DE must represent the reactions of the hinges a and γ respectively.

There is now no difficulty in completing the force diagram. EAO_1E (this way round) is the force triangle for the joint eao_1, ABO_2O_1A (this way round) is the force polygon for the joint abo_2o_1, EO_1O_2E for the joint eo_1o_2, $O_2BCO_3EO_2$, for the joint o_2bco_3e, etc.

EXAMPLES XVIII.

1. Three uniform rods KL, LM, MN, of lengths 33, 30, 33 inches respectively, and weighing 20 ounces, 16 ounces, 20 ounces respectively, are freely jointed together at L and M, and suspended from a fixed point H by means of four fine light strings HK, HL, HM, HN, of lengths 52, 25, 25, 52 inches respectively. Find the tensions of the strings.

2. Three uniform rods KL, LM, MN, of lengths 33, 30, 33 inches respectively, and weighing 20 ounces, 16 ounces, 20 ounces respectively, are freely jointed together at L and M, and suspended

from a fixed point H by means of three fine light strings HK, HF, HN, of lengths 52, 20, 52 inches respectively, the point F being the middle point of LM. Find the tensions of the strings and the actions at the hinges L and M.

3. Three uniform rods KL, LM, MN, of lengths 15, 14, 15 inches respectively, and weighing 3, 5, 3 pounds respectively, are freely jointed together at L and M, and suspended from a fixed point H by means of four fine light strings HK, HL, HM, HN, of lengths 20, 25, 25, 20 inches respectively. Find the tensions of the strings. .

4. Two uniform beams HL, KL, each of length 13 feet, and weighing 15 pounds, are freely jointed together at L, and supported from a fixed point M by means of two fine light strings MH, MK, each of length 15 feet. A fine light rod RS, of length 16 feet, is freely jointed to the two beams at the points R and S, which divide LH and LK respectively, each in the ratio 2 : 1. Find the thrust in the cross-rod, the tension of each string, and the action at the hinge.

5. Six equal uniform rods, each of weight W, are freely jointed at their extremities to form a regular hexagon $ABCDEF$. The rod AF is supported in a horizontal position, and distortion is prevented by a fine light rod, connecting the middle points of BC and DE. Find the thrust of the cross rod and the actions at the hinges.

6. Four equal uniform rods, each of weight W, are freely jointed together at their extremities to form a rhombus $HKLM$. The rod HK is supported in a horizontal position, and a fine light string, equal in length to one of the rods, connects the hinges M and K. Find the tension of the string and the actions at the hinges H and L.

7. Two equal uniform rods HL, KL, each of weight W, are freely jointed together at L, and supported from a fixed point M by means of three fine light strings HM, LM, KM, each equal in length to one of the rods. Find what load must be applied at K, so that LK may rest in a horizontal position; determine also the tensions of the strings.

8. Four equal uniform rods, each of weight W, are freely jointed together at their extremities to form a rhombus $HKLM$, which is supported at H. A fine light string, in length equal to half of one of the rods, connects the middle points of LM and HM. Find the tension of the string and the actions at the hinges K, L, M.

9. Work out the preceding example, supposing the string to be of length equal to three-quarters of one of the rods, and to be attached to the rods LM, HM at points which divide LM, HM each in the ratio $1:3$.

10. Three uniform rods FG, GH, HK, each weighing 15 pounds and of length 25 inches, are freely jointed together at G and H, and rest in a vertical plane upon a smooth horizontal table at F and K. Two fine light strings FH, GK, each of length 40 inches, help to support the framework, and a mass of 90 pounds is placed at the middle point of GH. Determine the tensions of the strings.

11. In the preceding example, determine the tensions of the strings when the mass of 90 pounds is placed at a distance of 6 inches from H.

12. Two equal uniform rods BC, CD, each of weight W, are freely jointed together at C, and connected with a fixed point A by means of three fine light strings AB, AC, AD, the first and the third being equal in length. The whole system rests with AC vertically downwards. Prove that if E is the point where BD intersects AC, and F the middle point of AE, then the weight W, the tension of the string AB, and the tension of AC are respectively proportional to $2.AC$, AB, $4.CF$.

13. Two uniform rods KL, LM, equal in length and weight, are freely jointed together at L, and connected with a fixed point H by means of two fine light strings KH, MH. Another fine light string connects H with L, and is of such a length that, when the strings are all tight, KL and LM are in one straight line, and KHM a right angle. The whole is allowed to rest in a vertical plane, being supported at H. Prove that, in the position of equilibrium, the tensions of the strings HK, HL, HM are proportional to HK, $2.HL$, HM respectively, the tension of the string HL being equal to the weight of either rod.

Prove also that, if the hinge at L is a separate piece to which the string HL is attached, the actions between the hinge and the rods KL, ML are respectively parallel to HM, HK. If equal loads are applied at K and M, prove that the tensions of the strings HK, HM are increased in the same ratio, the tension of HL remaining the same as before.

14. Four uniform rods are freely jointed together at their extremities to form a quadrilateral framework $ABCD$, the rods AB, AD being equal in length, and each of weight W, and the rods BC, CD equal, and each of weight w. A fine light string connects the hinges A and C, and the whole framework is supported at A. The straight line BD intersects AC at E; prove that, if T is the tension of the string,

$$T - w : W + w = CE : CA.$$

15. Four equal uniform rods, each of weight W, are freely jointed together at their extremities to form a rhombus $ABCD$, which is stiffened by a fine light rod connecting the hinges B and D. The whole is supported at A. Show that, if P is the thrust of the cross rod, $P : 2W = BD : AC$.

Show also that the action of the hinge C is $\frac{1}{4}P$.

16. Three equal uniform rods HK, KL, LM, each of weight W, are freely jointed together at K and L, and supported by four fine light strings FH, FK, GL, GM from two fixed points F and G, which are situated in a horizontal line at a distance apart equal to the length of one of the rods. The strings FH, FK are respectively equal to GM, GL, so that the system occupies a symmetrical position of equilibrium with KL horizontal. If S and T are the tensions of the strings FH and FK respectively, prove that

$$S : \tfrac{1}{2}W : T - W = HF : FK : KN,$$

N being the point where the straight line HM intersects FK.

17. Three uniform rods FG, GH, HK, of weights W, W', W'' respectively, the two FG, HK being equal in length, are freely jointed to one another at G and H, and laid in a vertical plane upon a smooth horizontal table, the extremities F and K being connected by a fine light string. The system takes up a symmetrical position of equilibrium with GH horizontal. Prove that,

if T is the tension of the string,

$$2T: W + W' = FL : LG,$$

where L is the point of FK vertically below G. Also, if GL be divided in M so that $GM : ML = W : W'$, prove that FM is parallel to the action at the hinge G.

18. If, in the preceding example, the string is attached to points which divide FG, KH each in the ratio $m : n$, prove that

$$2n . T : (m+n)(W + W') = FL : LG.$$

19. Four heavy uniform rods a, b, c, d are freely jointed at their extremities to form a quadrilateral framework. A fine light string connects the hinges ab and cd, and the framework rests in a vertical plane with the rod d held in a given position. Prove the following method for determining the tension of the string : Draw TBC vertically downwards, making TB to represent half the sum of the weights of a and b, and BC half the sum of the weights of b and c. Let the straight lines through B and C, parallel to the rods b and c respectively, meet in O. Draw TA and OA, parallel to the string and the rod a respectively, to meet in A. Then AT represents the tension of the string.

20. HKL is a fixed upright beam, the point H being at the bottom. Two horizontal uniform beams HM, KN are freely jointed to the fixed upright at the points H and K, and a vertical beam MN is freely jointed to the horizontal beams at the points M and N. The rectangular shape of $HMNK$ is preserved by means of a fine light string connecting M with L, the whole system resembling a gate. Show that, if T is the tension of the string, W_1 the weight of HM, w of MN, and W_2 of KN, then

$$T : \tfrac{1}{2}(W_1 + W_2) + w = LM : LH.$$

Show also that the actions at the hinges K and N are both vertical ; also that, if G be taken in MN such that

$$MG : HL = \tfrac{1}{2}W_1 : \tfrac{1}{2}(W_1 + W_2) + w,$$

then HG is the line of action of the constraint at H, which is of magnitude R such that

$$R : \tfrac{1}{2}W_1 = HG : GM.$$

CHAPTER XIX.

SOME MISCELLANEOUS PROBLEMS.

163. Ex. 1. *Two rods* AB, AC, *of given lengths and of no appreciable weight, are smoothly jointed together at A, and rest in a vertical plane upon a smooth horizontal plane at B and C, the points B and C being*

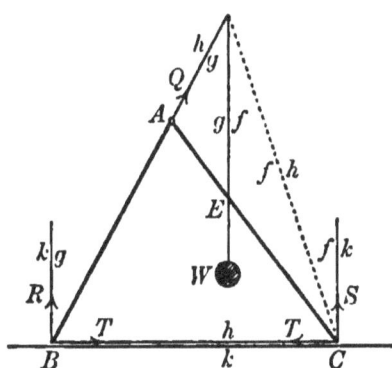
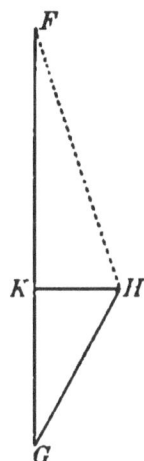

FIG. 181.   FIG. 181 a.

connected by a fine light string of given length. From a given point E of the rod AC is suspended a mass of given weight W. It is required to find the tension of the string, the action at A, and the pressures at B and C.

The data are sufficient to enable us to construct the space diagram. Let R and S be the reactions at B and C respectively, T the tension of the string, Q the mutual action of the hinge on either rod.

The rod AB is in equilibrium under the influence of forces acting only at its extremities. It is therefore in a state of direct compression or tension. Therefore the mutual action between the rods at A is in the line AB.

Thus the rod AC is in equilibrium under the influence of four forces whose lines of action are all known, but the magnitude of only one. We have, therefore, an example of Art. 103.

Draw FG vertically downwards, making it W units of length, and mark the vertical through E with the letters fg. Let the line BA and the horizontal and vertical through C be called gh, hk and kf respectively. Draw the straight line fh from C to the point of intersection of fg and gh. Draw FH, GH parallel to fh, gh respectively to meet in H, and HK horizontal to meet FG in K.

Then $FGHKF$ (this way round) is the force polygon for the rod AC, the forces Q, T and S being represented by GH, HK and KF respectively.

Also $GKHG$ (this way round) is the triangle of forces for the rod AB, so that GK represents R.

Otherwise: We may break up W into two forces acting at A and C, and proceed as in the preceding chapter.

164. Ex. 2. *AB and AC are two rods of no appreciable weight, smoothly jointed at A, and resting at B and C upon a smooth horizontal plane. A fine string*

*connects the point F of the rod AB with the point G
of the rod AC. The rod AB is loaded at D with a
mass of given weight W_1, and the rod AC is loaded at
E with a mass of given weight W_2. It is required to
find the tension of the string, the reactions at B and C,
and the action at A.*

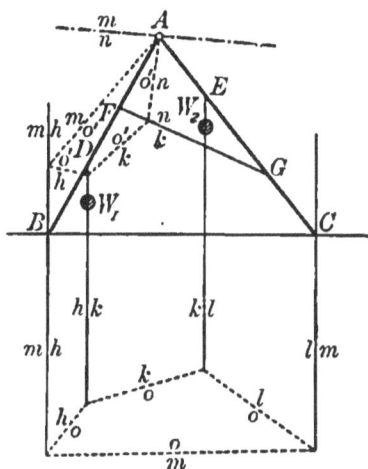

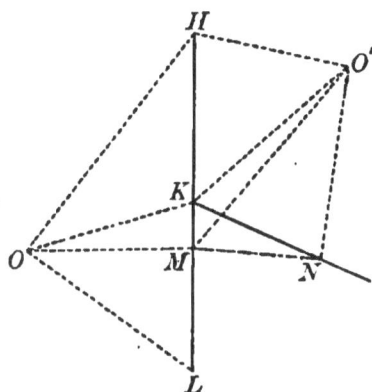

FIG. 182.   FIG. 182 a.

First, consider the equilibrium of the two rods together as one system. Let the verticals through
D, E, C, B be marked hk, kl, lm, mh respectively. Take
HK, KL to represent W_1 and W_2 respectively. Take any
pole O, and from any point in mh draw oh parallel to
OH. From the point of intersection of oh and hk draw
ok parallel to OK. From the point of intersection of
ok and kl draw ol parallel to OL. Draw the straight
line om, joining the point of intersection of ol and lm
with the point of intersection of oh and hm. This completes the funicular polygon. Draw OM parallel to om
to meet HL in M. Then LM, MH represent the reactions
at C and B respectively.

Now consider the equilibrium of the rod AB alone. The forces acting on it are the known forces along mh and hk, the tension T of the string in the direction FG, which we mark kn, and the unknown action at A in some line which will be denoted by nm. Draw KN parallel to kn. Then the force polygon will be $MHKNM$ (this way round), where at present the point N is unknown. The pole O now gives an awkward figure; we therefore take a new pole O'. Starting from A, we draw the strings $o'm$, $o'h$, $o'k$ of the funicular polygon, which we complete by drawing the string $o'n$ from A to the intersection of $o'k$ and kn. Draw $O'N$ parallel to $o'n$ to meet KN in N. Then, joining MN, we complete the force polygon. KN represents the tension of the string, and NM the action of the hinge upon the rod AB.

The force polygon for the rod AC is $KLMNK$ (this way round).

165. *Ex. 3. Four rods, of no appreciable weight, are freely jointed together at their extremities to form the quadrilateral ABCD. The framework is stiffened by another light rod smoothly hinged to the point E of the rod AB, and to the point F of the rod AD. Equal forces P are applied at A and C in opposite directions along the line AC. It is required to find the stress in the cross rod, and the actions at the hinges.*

The three rods BC, CD, EF are in equilibrium under the influence of forces acting only at their extremities. They are therefore in a state of direct compression or tension. Thus the actions at B and D are in the lines BC, CD respectively, and the actions at E and F are both in the line EF.

Let EF produced both ways meet CB produced and CD produced in L and M respectively. Then, considering the equilibrium of the rod AB, we see that the action of the hinge A upon this rod is in the line AL. Similarly the action of the hinge A upon the rod AD is in the line AM.

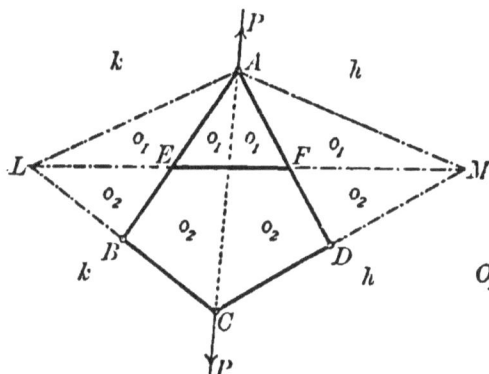
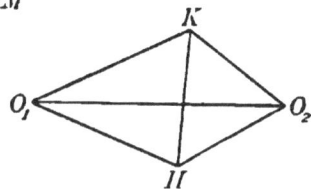

FIG. 183.   FIG. 183 a.

Now let the space inside the triangle LMA be denoted by o_1, and the space within the triangle LMC by o_2. Also let the space outside the figure and to the right of AC be denoted by h, and the space outside the figure and to the left of AC by k.

For the equilibrium of the joint C, we have the triangle KHO_2K (this way round), in which KH represents P, and HO_2, KO_2 are parallel to the lines ho_2, ko_2 respectively.

For the equilibrium of the rod AB we have the triangle KO_2O_1K (this way round), in which O_2O_1, KO_1 are parallel to the lines o_2o_1, ko_1 respectively.

Hence, joining O_1H, the triangle HO_1O_2H must be the triangle of forces for the rod AD, so that O_1H must be parallel to the line o_1h.

Also, if we consider the equilibrium of the hinge alone at A, the triangle of forces is HKO_1H (this way round).

The student will see that the force diagram is the same as for the equilibrium of rods LA, AM, MC, CL, stiffened by a rod LM, and under the same external forces P, applied at A and C.

166. Ex. 4. *Four rods, of no appreciable weight, are freely jointed together at their extremities to form the quadrilateral ABCD. The framework is stiffened by another light rod, connecting the hinge D with a point E of the rod BC. Equal forces P are applied at A and C in opposite directions along the line AC. It is required to determine the stress in the cross rod.*

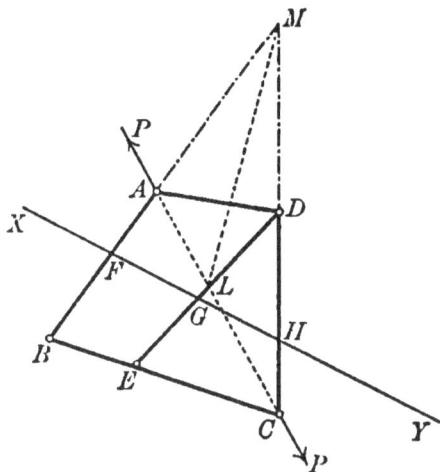
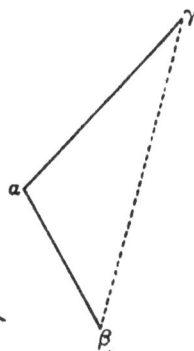

FIG. 184.   FIG. 184 a.

This, of course, can be worked out in the same way as the preceding problem, of which it is a particular case, but as we require only the stress in the cross rod, we may proceed as follows:

Draw a line XY intersecting the three rods AB, ED, DC in F, G, H respectively, and consider the equilibrium of the portion $FBECHG$ as a rigid body. The three rods intersected are all in a state of direct compression or tension.

The external forces acting upon this part of the framework are: the force P at C, and the actions at F, G, H, which are in the lines AB, ED, DC respectively. The force P and the action at G must balance the actions at F and H. Find L and M, the points of intersection of AC, ED and BA, CD respectively. Then LM must be the line of action of the resultant of the force P and the action at G.

Hence, drawing $\alpha\beta$ to represent the force P acting at C, let straight lines through α and β, parallel to ED and LM respectively, meet in γ. Then $\gamma\alpha$ represents the action of the portion DG upon the portion GE, thus determining the stress in the rod ED, and showing whether the rod is a *tie* or a *strut*.

The method is applicable even if the point M is inaccessible, for we only require the direction of LM, and we can draw the straight line from L towards the inaccessible point of intersection of AB and CD.

In applying this method care must be taken to intersect only such rods as are acted upon by forces at their extremities only.

EXAMPLES XIX.

1. Two rods AB and AC, each of length 2 feet 3 inches and of no appreciable weight, are smoothly jointed together at A, and placed, in a vertical plane, with B and C on a smooth horizontal plane. The points B and C are connected by a fine string,

of length 2 feet 8 inches, and from a point E of the rod AC, distant 1 foot from C, is suspended a mass of 30 pounds. Find the tension of the string.

2. AB and AC are two equal rods, of no appreciable weight, smoothly jointed together at A, and resting in a vertical plane upon a smooth horizontal plane BC. D is a point in AB such that $AD = \frac{1}{3}AB$, and E and F are points in AC such that $AE = EF = FC$. A fine string connects D with F, and is of such a length that the angle A is 60°. Find the tension of the string when a mass of 60 pounds is suspended from E. Determine, also, the magnitude of the action at A.

3. AB and CD are two rods, each of length 4 feet and of no appreciable weight, freely jointed together at C, the middle point of AB. A fine string, 5 feet long, connects A and D, and the whole rests in a vertical plane upon a smooth horizontal plane AD, a mass of 100 lbs. being suspended from B. Find the tension of the string AD and the thrust in the rod CD.

4. Solve Ex. 5, Art. 160, by the method explained in Art. 164.

5. $ABCD$ is a rhombus formed by four rods, of no appreciable weight, freely jointed together, and the figure is stiffened by another rod, of inappreciable weight and of half the length of each side of the rhombus, jointed to the middle points of AB and AD. If this framework is suspended from A, and a mass of 100 pounds attached to it at C, find the thrust of the cross rod.

6. $ABCD$ is a rhombus formed of four rods, of no appreciable weight, loosely jointed together, so that ABD and BCD are equilateral triangles. The framework is stiffened by another light rod DE, connecting D with the middle point of BC. If this framework is suspended from A, and a mass of weight W attached to it at C, find the thrust of the cross rod.

7. In the example of Art. 163, the vertical through E meets BC in X, and the straight line through C, drawn perpendicular to AB, meets EX in Y. Prove that

$$W : T : R : S : Q = BC : YX : CX : XB : CY.$$

8. In the preceding example, suppose that the weight of the mass is given, but that the point E may be anywhere in AC. Show that the tension of the string varies as CE.

9. Two rods AB and AC, of no appreciable weight, are smoothly jointed together at A, and rest in a vertical plane upon a smooth horizontal plane at B and C, the point B being connected with a point D in AC by a fine string. From a point E of the rod AC is suspended a heavy body. The vertical through E meets BC in X, and straight lines through X and C, perpendicular to BD and BA respectively, meet in Y. Prove that the weight of the body, the tension of the string, and the stress in AB are proportional to BC, XY, CY respectively.

10. Four rods, of no appreciable weight, are freely jointed together at their extremities to form a quadrilateral $ABCD$, such that AB is parallel to DC. The framework is stiffened by another light rod connecting the hinge D with a point E of the rod BC. Equal forces P are applied at A and C in opposite directions in the line AC, which intersects ED at the point F. Prove that, if T is the stress in the rod ED, then

$$T : P = DF : FC.$$

CHAPTER XX.

FRICTION.

167. SUPPOSE that a rigid body rests in equilibrium against a rough surface *at one point*, being acted upon by a given system of forces in one plane, in addition to the resistance of the surface.

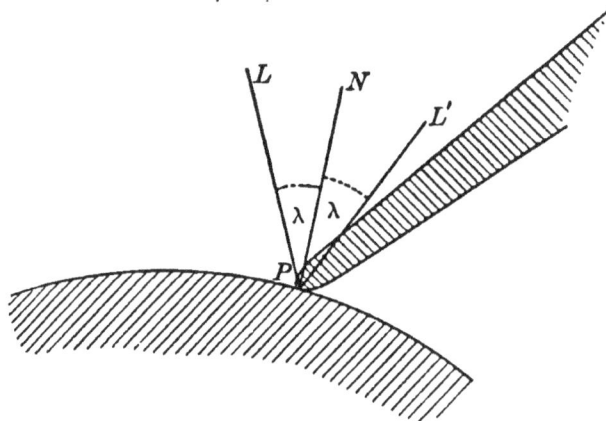

FIG. 185.

Let the body rest against the surface at *P*.

Draw *PN*, the normal, away from the surface at *P*.

Make angle *NPL*=angle of friction (λ)=angle *NPL'*.

Then the total resistance at *P* can be of any magnitude, but must act in a direction intermediate to *PL* and *PL'*. For equilibrium, then, it is necessary and sufficient that the resultant of the given system of forces should pass through P and be in a direction intermediate to LP and $L'P$.

168. For example, suppose it is required to support a given heavy rod in a given position with one end resting against a given rough inclined plane, by applying at some point of the rod a force in a given direction.

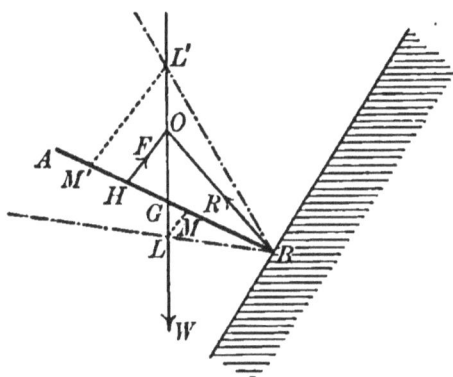

FIG. 186.   FIG. 186 a.

Let AB be the rod resting against the plane at B. Draw BL and BL', making the angle of friction on either side with the normal drawn away from the plane at B. Let the vertical through G, the centre of gravity of the rod, meet BL, BL' in L and L' respectively.

Draw LM and $L'M'$ each parallel to the given direction of the applied force, to meet AB in M and M' respectively. Then the force must be applied at some point between M and M'.

For, the only external forces acting on the rod are its weight (W), the total resistance of the plane (R), and the applied force (F). The lines of action of the first two of these forces intersect at a point between L and L'. Hence the line of action of F must pass through some point between L and L'. Thus, for equilibrium, it is necessary that the force F should be of suitable magnitude, and should be applied at some point H between M and M'.

For any given point H between M and M' we can determine the magnitude of the applied force, and the total resistance at B. Draw HO in the given direction of the applied force, to meet LL' in O. Then BO must be the line of action of R.

Draw $\alpha\beta$ to represent the weight of the rod, and let straight lines be drawn through α and β parallel to OB and HO respectively, to meet in γ. Then $\beta\gamma$ represents the force F, and $\gamma\alpha$ the force R.

It will be seen that we have taken the inclination of the plane as greater than the angle of friction. If the inclination is less than the angle of friction, the vertical through G will not meet BL' above B. In this case the point O may have any position above L.

169. In the preceding article, we have said that, for equilibrium, it is necessary that the force F should be applied at some point H between M and M'. This, however, is not the only condition of equilibrium. We have to take into account another consideration;—the resistance at B must be in direction BO and not OB. It is necessary, therefore, to examine the direction arrows in the triangle $\alpha\beta\gamma$, and see that $\gamma\alpha$ indicates a *push* at B and not a *pull*.

For instance, suppose that M' lies in AB produced, the inclination of the plane being greater than the angle of friction. It will be found that, if H is taken anywhere except between B and M', the triangle of forces indicates that a *pull* would be required at B; and, as the force F cannot be applied at a point in AB produced, it follows that in this case the rod cannot be supported.

170. *A rigid rod AB, whose centre of gravity is at G, rests against a rough horizontal plane at. A, and a rough vertical wall at B. The angles of friction between the rod and the ground, and between the rod and the wall are λ and λ' respectively. It is required to consider the conditions of equilibrium when any system of forces is applied to the rod.*

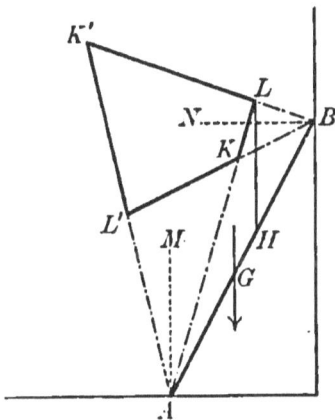

FIG. 187.

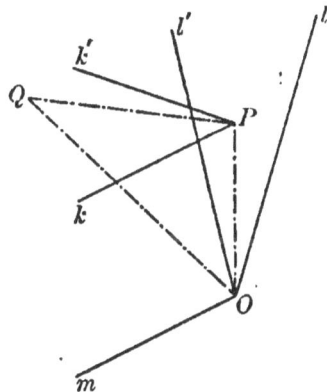

FIG. 187 a.

Let AM, BN be drawn normals to the ground and wall respectively. Draw AK, AK', each making with AM on opposite sides an angle λ, and draw BL, BL', each making with BN on opposite sides an angle λ'.

The total resistance at A may be of any magnitude, but must act within the angle KAK', and the total resistance at B may be of any magnitude, but must act within the angle *LBL'*. Hence the lines of action of the total resistances at *A* and *B* intersect at some point within the quadrilateral *KLK'L'*. If it is possible, these resistances will be so adjusted as to produce equilibrium. Hence, for equilibrium, it is necessary and sufficient that the resultant of all the other forces acting on the rod should be a force whose line of action intersects the quadrilateral *KLK'L'*, and whose direction is such as to produce *pressures* at *A* and *B*.

The limits to the direction of this resultant force are determined in this way : Let *OP* represent the resistance at *A*, and *PQ* the resistance at *B*. Then *QO* must represent the resultant of all the other forces acting on the rod. The point *P* lies anywhere within the angle *lOl'*, where *Ol, Ol'* are drawn in the directions of *AL, AL'* respectively; and the point *Q* lies anywhere within the angle *kPk'*, where *Pk, Pk'* are drawn in the directions of *BK, BK'* respectively. Hence, drawing *Om* in the direction of *Pk*, we see that the only limitation upon the position of the point *Q* is that it must be somewhere within the angle *mOl*. Hence the resultant of all the forces acting on the rod, other than the resistances at its extremities, must be intermediate in direction between *KA* and *KB*.

171. If the only forces acting on the rod are its weight and the resistances at its extremities, then, for equilibrium, it is necessary and sufficient that the vertical through *G* should intersect the area *KLK'L'*. If *G* is vertically below *L*, the equilibrium is limiting, and the actions at *A* and *B* are then along *AL* and *BL* respectively, and are determinate. In other cases these resistances cannot be found.

Fig. 187 represents a possible position of equilibrium, as the vertical through G intersects the area $KLK'L'$.

Let w be the weight of the rod, and let the vertical through L meet AB in H. Then it is clear that if any load be applied to the rod at a point between A and H, equilibrium will not be disturbed. If, however, a load W be applied at a point J between H and B, equilibrium will, or will not, be disturbed, according as the resultant of w and W acts above or below H.

172. Let the point J be given, and suppose it is required to determine the greatest value of W consistent with equilibrium. We may *measure GH* and HJ, and find W from the equation

$$\frac{W}{w} = \frac{GH}{HJ}.$$

Or, we may obtain the same result as follows:

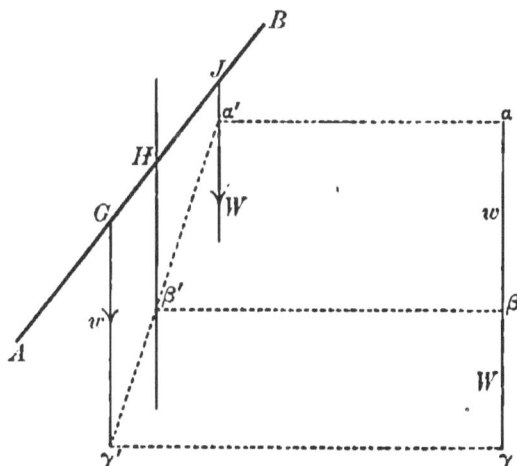

FIG. 188.

Draw $\alpha\beta$ to represent w, and let two parallel lines through α and β meet the verticals through J and H in α' and β' respectively. Produce $\alpha'\beta'$ to meet the vertical through G in γ', and draw $\gamma'\gamma$ parallel to $a'a$ to meet $a\beta$ produced in γ. Then $\beta\gamma$ represents W.

For any smaller load applied at J, the equilibrium will not be broken, and for any larger load equilibrium will be impossible.

173. Again, let W be given, and let it be required to find the extreme position of J consistent with equilibrium.

Here we can *measure* GH, and find HJ from the equation

$$\frac{HJ}{GH} = \frac{w}{W}.$$

Or, we may obtain the same result as follows:

Draw $a\beta$, $\beta\gamma$ to represent w and W respectively, and let two parallel lines through β and γ meet the verticals through H and G in β' and γ' respectively. Produce $\gamma'\beta'$ to meet the parallel through a in a', and draw $a'J$ in a vertical direction to meet AB in J.

If the load be placed below J, the equilibrium will not be broken; if above J, equilibrium will be impossible.

174. Suppose that a body rests in equilibrium against a rough inclined plane, a flat portion of the surface of the body being in contact with the plane. Let it be acted upon by a given system of forces, all situated in the vertical plane through a line of greatest slope of the inclined plane, these forces being in addition to the resistances of the inclined plane, which prevent the body from either penetrating, or slipping along, the inclined plane.

The resultant resistance of the plane now acts at some point within the portion of the plane in contact with the body, and makes an angle with the normal at that point not greater than the angle of friction.

Hence, for equilibrium, it is necessary and sufficient that the resultant of the given system of forces should act *towards* the plane, and intersect the body at a point within the extreme limits of the surface in contact, and that it should not make with the normal to the plane an angle greater than the angle of friction.

If the resultant of the given system of forces acts along a line outside the figure formed by a string drawn tightly round the portion of the body in contact with the plane, the tendency of the forces is to overturn the body. If the resultant of the given system of forces makes an angle with the normal greater than the angle of friction, the tendency is to make the body slip along the plane.

175. For instance, consider the following problem:

A lamina of given shape, and of weight W, rests with a straight edge AB in contact with a rough inclined plane, B being above A. It is supported by a force

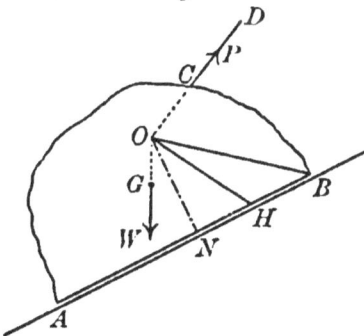
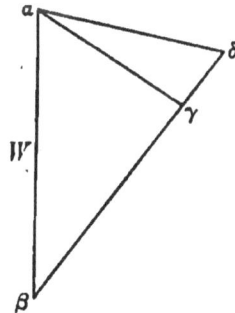

FIG. 189.   FIG. 189 a.

applied in a given direction at a given point C. This force is gradually increased until motion ensues. It is required to find whether the lamina slips or topples over.

FRICTION.

Let CD be the line of action of the force applied at C in a direction tending to move the body *up* the plane. Let G be the centre of gravity of the body, and let the vertical through G meet DC produced (if necessary) in O. Draw ON normal to the plane and OH, making with ON the angle of friction, such that NH is *up* the plane.

If H lies *within* AB, the body will slide; if *without*, the body will topple over round B.

For, draw $a\beta$ to represent W, and let straight lines be drawn through a parallel to OH, OB to meet a line through β parallel to CD in γ and δ respectively. If the force applied at C is increased until it has a value represented by $\beta\gamma$, the body will be just on the point of sliding; if it is increased until it has a value represented by $\beta\delta$, the body will be just on the point of toppling over. To ascertain what *actually* takes place, we have merely to see which is the smaller, $\beta\gamma$ or $\beta\delta$; and, clearly, $\beta\gamma$ will be less or greater than $\beta\delta$ according as H lies within or without AB.

176. Ex. 1. *A and B are two fixed pegs, B being at a higher level than A, and a heavy rod rests on B and passes under A. The angle of friction between the rod and the pegs being the same for both, it is required to determine the conditions of equilibrium.*

Let a be the inclination of AB to the horizon, λ the angle of friction.

Draw AM, nBN, the normals at A and B respectively. Make angle $KAM = \lambda =$ angle MAK'. Also draw LBl, $L'Bl'$, making each an angle λ with nBN, and let Bl meet AK' in H, as in the figure. Draw BFG vertically downwards, meeting AK' in F. Then angle $GBn = a$.

The resistance at A acts in a direction intermediate to AK and AK'. That at B acts in a direction intermediate to lB and $l'B$. Hence the lines of action of these two resistances intersect at a point within the area lHK', and equilibrium is possible only when the line of action of the resultant weight of the rod intersects the same area.

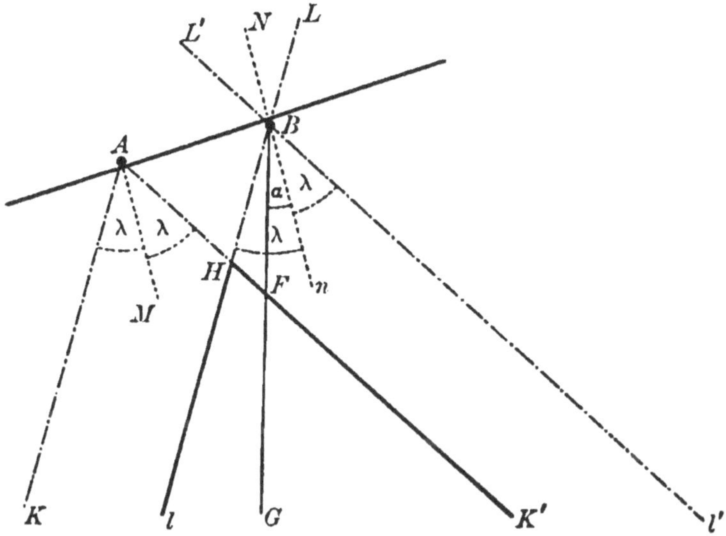

FIG. 190.

I. Let a be $< \lambda$. Then BG is within the angle lBn, so that FG divides the area lHK' into two parts. The line of action of the weight of the rod must not fall within the area $lHFG$, otherwise it would be necessary for the peg at A to *pull* instead of *press* the rod. Hence, for equilibrium, it is necessary and sufficient that the vertical through the centre of gravity of the rod should intersect the area GFK'. Thus the rod will rest with its centre of gravity anywhere above B.

II. Let a be $> \lambda$. Then BG is without the angle lBn. In this case equilibrium will always exist so long as the line of action of the resultant weight of the rod intersects the area lHK'.

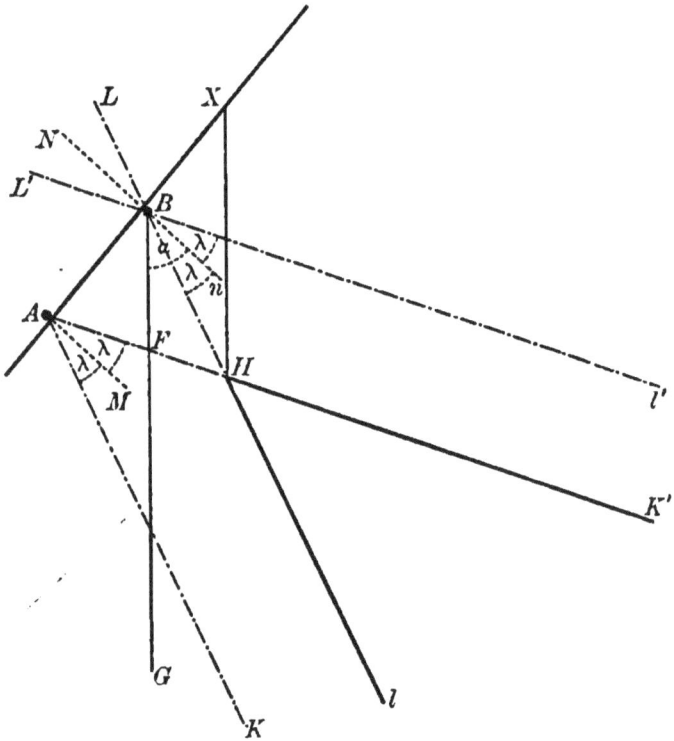

FIG. 191.

Hence, if the vertical through H meets AB in X, equilibrium will exist so long as the centre of gravity of the rod is not below X.

177. *Ex.* 2. *A uniform heavy beam rests with one end against a vertical wall and the other on the ground, being inclined to the wall at an angle of* 45°. *Compare the least horizontal forces which, applied to the foot of the*

*beam, will move it towards or from the foot of the wall,
the coefficients of friction being* $\frac{1}{4}$ *for each end of the
beam.*

Draw AB to represent the beam inclined at an angle
of 45° to AD, which represents the ground, and to
BD, which represents the wall. (For figure, see next
page.)

Let the normals at A and B intersect at N, and draw
AH and AK, intersecting BN in E and F respectively,
where $EN = \frac{1}{4}AN = NF$. Then

angle HAN = angle of friction = angle NAK.

Similarly, make

angle HBN = angle of friction = angle NBK.

Then the lines of action of the total resistances at
A and B intersect at some point within the shaded area
of the figure.

Let the vertical through the middle point of the beam
meet the horizontal through A in O. Let W be the
weight of the beam, and suppose it is at rest when a
force X is applied at A in a horizontal direction towards
the wall.

The line of action of the resultant of X and W
passes through O. If it is possible for the resistances at
A and B to adjust themselves so as to balance the
resultant of X and W, they will do so. Hence, for
equilibrium, it is necessary and sufficient that the line
of action of the resultant of X and W should pass
through the shaded area. Thus the line of action of
the resultant of X and W must lie between the positions
HO and KO. If it is beyond these limits, the beam
slips; if it is along HO, the beam is on the point of

slipping *towards* the wall; if along KO, the beam is on the point of slipping *away from* the wall.

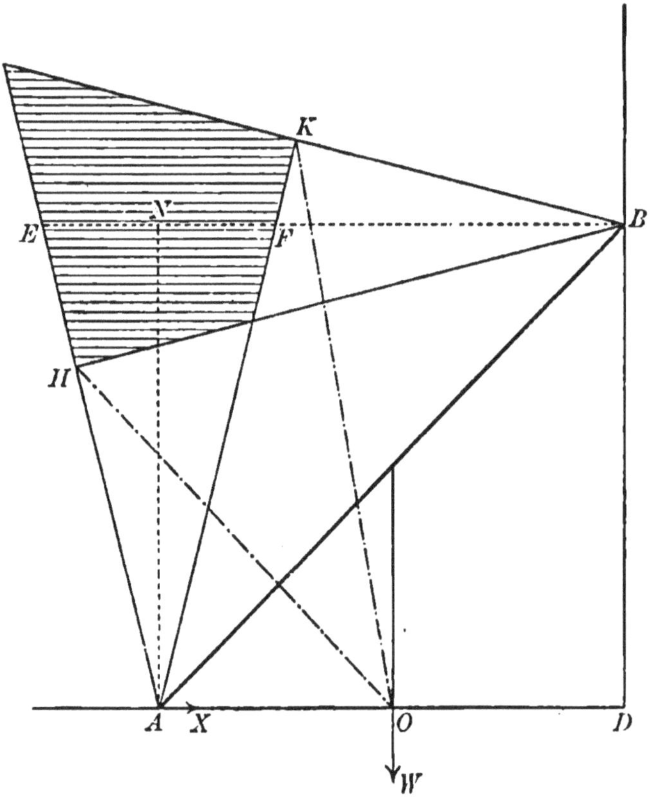

FIG. 192.

Draw $a\beta$ vertically downwards to represent W, and draw straight lines through a. parallel to HO, KO to meet the horizontal through β in γ and δ respectively. Then $\beta\gamma$, $\beta\delta$ represent the greatest and least values of X consistent with equilibrium.

On measurement, we find that $\beta\gamma$ is $5\frac{1}{2}$ times as long as $\beta\delta$, therefore the force necessary to move the beam towards the wall is $5\frac{1}{2}$ times as great as the force necessary to prevent it from slipping away from the wall.

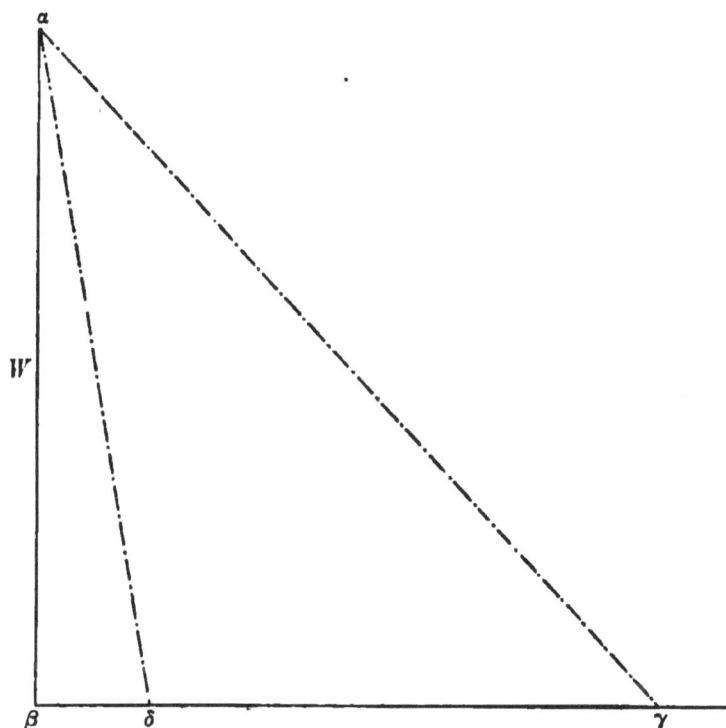

FIG. 192 a,

178. Ex. 3. *A ladder is placed with one end on a rough horizontal plane and the other against a rough vertical wall; find, by geometrical construction, the limiting position of equilibrium, being given the co-efficients of friction and the centre of gravity of the ladder.* •

If an additional load be placed at any point on the ladder, in this limiting position, find whether the equilibrium will be disturbed or not.

We will assume a position of limiting equilibrium,

and postpone for the present any attempt to draw the figure to scale.

Let AB represent the ladder, resting against the ground AD at A and the wall BD at B. Let G be the centre of gravity of the ladder, μ and μ' the co-efficients of friction for the ladder and ground and for the ladder and wall respectively. Then μ and μ' are known fractions, and the lengths of AG and GB are known.

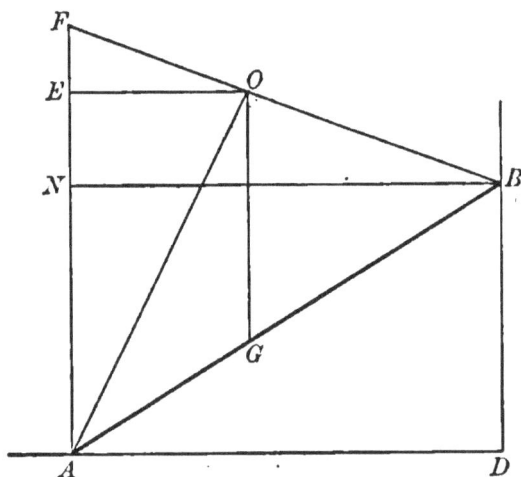

FIG. 193.

Let the normals at A and B intersect at N. Suppose that NAO is taken equal to the angle of friction for the end A, and NBO the angle of friction for the end B, the first angle being measured•from NA towards the wall and the second upwards from NB. Then, as the ladder is in *limiting* equilibrium, the vertical through G must pass through O. Let the vertical through A meet the horizontal through O in E and BO produced in F.

Then EAO, EOF are the angles of friction for the ends A and B respectively.

$$\therefore \quad EO = \mu . EA \text{ and } EF = \mu' . EO.$$

Also $\qquad FO : OB = AG : GB.$

Hence we have the following method for constructing the figure:

Take any straight line EA vertically downwards. Through E draw EO in a horizontal direction equal to $\mu . EA$, and produce AE to F, making $EF = \mu' . EO$.

Produce FO to B, making the ratio $FO : OB$ equal to the known ratio $AG : GB$. Then AB may be taken to represent the ladder.

In the construction indicated above, EA is taken of any suitable length, without reference to a scale, and the line AB is constructed, its length depending upon the length chosen of EA. We can then choose our scale so that AB may represent the known length of the ladder; or, if this is inconvenient, we can draw another figure, similar to the figure obtained, commencing with AB, which is first drawn to scale.

If an additional load be placed on the ladder between A and G, the resultant weight of ladder and load will act along a vertical line to the left of O, and therefore the equilibrium will not be disturbed.

If, however, an additional load be placed on the ladder between G and B, the resultant weight will act along a vertical line to the right of O, and therefore equilibrium will become impossible.

179. Ex. 4. *The uniform square lamina $ABCD$ rests vertically with the side BC upon a horizontal plane, coefficient of friction $\frac{1}{2}$, and has a fine string attached at D and passing over a small smooth peg at the point*

E in BA produced till EA is equal to AB. If the string be pulled, find the greatest force which can be applied, consistent with equilibrium, and whether the initial motion of the lamina will be tilting or sliding.

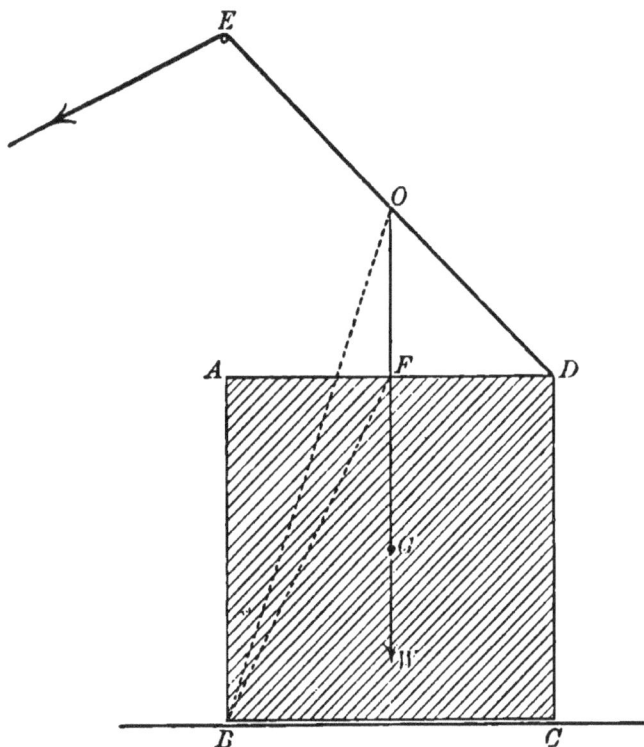

FIG. 194.

Let W be the weight of the lamina, and G its centre of gravity. Let the vertical through G meet AD in F and DE in O; join BF and BO.

The point F bisects AD, and therefore $AF = \frac{1}{2}AB$. Hence ABF is the angle of friction.

Take $\alpha\beta$ vertically downwards to represent W, and draw $\beta\gamma$, a straight line of unlimited length, parallel to DE.

If the point γ be taken so that $a\gamma$ is parallel to FB, the straight line $\beta\gamma$ will represent the pull of the string when the lamina is on the point of sliding. If γ be taken so that $a\gamma$ is parallel to OB, then $\beta\gamma$ will represent the pull of the string when the lamina is on the point of tilting round B. As FB is inclined to the vertical at a greater angle than OB, it follows that the second of these alternatives gives the shorter value of $\beta\gamma$. Hence the initial motion is one of *tilting*.

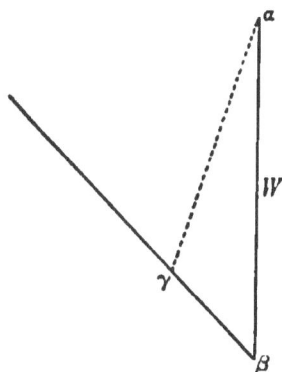

FIG. 194 a.

Also, drawing $a\gamma$ parallel to OB, we find, on measurement, that $\beta\gamma = \cdot 35$ of $a\beta$. Therefore, the greatest value of the applied force, consistent with equilibrium, is $\cdot 35\ W$.

EXAMPLES XX.

1. A uniform beam AB, whose length is $12\frac{1}{2}$ feet, rests with one extremity A on a rough horizontal plane AC, and is kept from falling forwards by a fine cord BC, 20 feet long, whose extremity is attached to a fixed point C in the plane, directly behind the beam. If the beam is on the point of slipping when $AC=AB$, find the coefficient of friction.

D.S.   X

2. A uniform beam of weight W, laid on a horizontal plane, can be just moved in its own direction by pushing it with a horizontal force ·58 W. Find the least force which can move it in its own direction, and determine the direction of this force.

If the beam be pulled by a gradually increasing force applied by a fine string attached at one end A, determine the least inclination of the string to the horizon, in order that A may be raised from the ground.

3. A uniform ladder 70 feet long is equally inclined to a vertical wall and the horizontal ground, both rough ; a man, ascending the ladder, weighs with his burden 2 cwt., and the ladder weighs 4 cwt.; how far up the ladder can the man ascend before it slips, the coefficient of friction for the ladder and wall being $\frac{1}{3}$, and for the ladder and ground $\frac{1}{2}$?

4. A ladder AB, 15 feet long, rests against the ground at A and against a rough vertical wall at B, the coefficients of friction at A and B being $\frac{3}{4}$ and $\frac{1}{2}$, respectively ; the centre of gravity, G, is 6 feet from A : find the inclination to the horizon at which the ladder will be just about to slip.

5. If the ladder of the preceding question is placed so that the distance of A from the wall is twice as great as the distance of B above the ground, and a boy, whose mass is one-fifth of that of the ladder, ascends it in this position, how far will he be able to go before the ladder begins to slip?

6. A uniform ladder rests between a vertical wall and the horizontal ground, both rough; if the coefficient of friction for the ladder and wall is $\frac{1}{3}$, and for the ladder and ground $\frac{2}{7}$, find the angle which the ladder makes with the ground when it just begins to slide.

7. A rectangular block $ABCD$, whose height is double its base, stands with its base AD on a rough floor, coefficient of friction $\frac{1}{5}$. If it is pulled by a horizontal force at C till motion ensues, determine whether it will slip on the floor or begin to turn over round D.

8. A uniform cubical block is sustained on a rough inclined plane by a fine string, which is parallel to the plane, and is

attached to the middle point of the upper edge of the cube, which is horizontal. The string lies in the vertical plane which contains the centre of the cube, and which is perpendicular to the inclined plane. The angle of friction being given, show how to determine the greatest inclination of the plane consistent with equilibrium.

Show that the greatest inclination is such that the ratio of the height of the plane to its base is $1+2\mu : 1$, where μ is the coefficient of friction.

9. A heavy rigid beam rests against a rough horizontal plane and against a rough vertical wall, the vertical plane through the beam being at right angles to the wall. Show that if the beam is inclined to the vertical at an angle less than the angle of friction for the beam and the ground, equilibrium cannot be broken, whatever loads be applied to the beam.

10. A heavy ladder is placed in a given position between a vertical wall and the horizontal ground, both being equally rough ; a workman of given weight ascends the ladder with a given load, show how to determine by a geometrical construction whether the ladder will slip.

11. In Example 3, Art. 178, show that, if the ladder is uniform, $BD : DA = 1 - \mu\mu' : 2\mu$.

If, in addition, $\mu = \mu'$, show that the limiting inclination of the ladder to the vertical is twice the angle of friction.

12. A ladder AB, whose centre of gravity is at G, rests against a rough horizontal plane at A, and a rough vertical wall at B, the coefficients of friction for the ground and the wall being μ and μ' respectively. Show that, if AG is less than $\mu\mu'$. BG, the ladder will rest at any inclination to the wall.

13. A uniform ladder rests in limiting equilibrium against a rough vertical wall and rough horizontal ground. Show that a man can ascend to the top of the ladder, while in this position, provided that a man, of not less weight than himself, stands on the ladder at the bottom.

14. A ladder, loaded in any manner, rests against a rough vertical wall and rough horizontal ground, being prevented from

slipping by means of a force applied at the foot of the ladder. Show that slipping is most easily prevented by applying the force downwards at an inclination to the ground equal to the angle of friction between the ladder and the ground.

15. A heavy rectangular block *ABCD* rests with *AB* on the ground ; a fine string is attached to the corner *C*, and pulled round a fixed smooth peg *P*, situated vertically over *D*, till motion ensues. All particulars being given, obtain a geometrical construction for determining the least height for the peg *P*, if the block is to begin to revolve round *A*, without slipping along the ground.

ANSWERS TO THE EXAMPLES.

EXAMPLES I.

1. (i.) 7 pounds' weight ; $21\frac{3}{4}°$;
   (ii.) 7 pounds' weight ; $38\frac{1}{4}°$;
   (iii.) 7 pounds' weight ; $21\frac{3}{4}°$.

2. 3·46 pounds' weight, parallel to AB, through the intersection of FB and AE.

3. 3 pounds' weight, acting through C in direction AB. 3 pounds' weight, acting through C in direction perpendicular to CA.

4. 56 pounds' weight, acting along the perpendicular from A upon BC.

5. 264·6 pounds' weight, acting at an angle of 19° with the first force.

6. 22·69 pounds' weight, acting at an angle of $73\frac{1}{2}°$ with the first force.

7. The other component is a force of 7 pounds' weight, acting at an angle of $81\frac{3}{4}°$ with the given component.

8. $47\frac{1}{2}°$.

9. 29·8, 45·9 pounds' weight.

10. 3 or 5 pounds' weight.

11. 11 or 24 pounds' weight.

12. An equal force, acting along the bisector of the angle between the given forces.

30. $OB : OA$.

EXAMPLES II.

1. 173·2, 200, 173·2 pounds' weight. 360·6 pounds' weight, acting at an angle of 46° with AB.

2. 0.

3. 0.

4. A force represented by twice AC.

5. 7·55 P, within the angle BOC, in a direction $53\frac{1}{2}°$ with OC.

6. 5.

7. 2·75 pounds' weight in direction BA.

8. 3·73, in the direction of the force 2.

9. $P=3·80$, $Q=6·12$.

10. 2·64, in a direction making 106° with the force 3.

11. $P=1·732=Q$.

12. 7 pounds' weight, acting at an angle of $21\frac{3}{4}°$ with the middle force.

13. 7 pounds' weight, acting at an angle of $38\frac{1}{4}°$ with the first force.

14. 17 pounds' weight, acting at an angle of 28° with AD.

15. 1·27.

16. 4·09 pounds' weight, acting at an angle of 83° with BC.

17. 7 pounds' weight, acting along OD.

18. 105 pounds' weight, acting along OF, where F is in AB at a distance of 3 inches from B.

19. 42·77 pounds' weight, acting at an angle of $70\frac{1}{2}°$ with the first force.

20. 7·89.

21. 13·2 pounds' weight ; $45\frac{3}{4}°$.

22. 1·155.

23. (i.) The forces cannot be arranged so as to produce equilibrium.

    (ii.) 135°;   (iii.) 90°;   (iv.) 120°;

    (v.) 120°;   (vi.) 60°.

EXAMPLES III.

1. 21·5, 6·4 pounds' weight.
2. 11 : 7.
3. 12, 16 pounds' weight.
4. 150 pounds' weight each.
5. 78, 50 pounds' weight.
6. 253, 91 pounds' weight.
7. 34, 20 pounds' weight.
8. $P=60$ pounds' weight ; the tension of each portion of the string $=50$ pounds' weight.
9. Tension in $BC=25$ pounds' weight$=$tension in CA. Tension in $AB=16·21$ pounds' weight. Each of the applied forces$=39·2$ pounds' weight.
10. Force applied at $B=51$ pounds' weight. Tensions in BC, CA, $AB=40$, 68, 13 pounds' weight respectively. Action at $A=75$ pounds' weight, in direction perpendicular to BC.
11. 33 pounds' weight.
12. 35 pounds' weight, 13 inches.

EXAMPLES IV.

1. 45 pounds' weight, 75 pounds' weight, 75 pounds' weight in direction AB. 120 pounds. $19\frac{1}{2}°$ or 53° downwards from the horizontal.
2. In AC, a thrust equal to the weight of 2·63 cwt.; and in BC, a tie equal to the weight of 1·75 cwt. A force equal to the weight of 9 cwt., acting in a direction perpendicular to AC and downwards. 15·2 cwt.
3. 64·66 pounds' weight.

4. In each side of the rhombus, tension of 84·66 pounds' weight. In the cross rod, thrust of 127·0 pounds' weight.

5. 76·92 pounds' weight.

6. 31·6 pounds' weight.

7. AC is vertically downwards; the tensions of the sides of the parallelogram are 80, 60, 80, 60 pounds' weight respectively; the thrust in BD is 124 pounds' weight.

8. AC is vertically downwards; the tensions of the sides of the parallelogram are 40, 30, 40, 30 pounds' weight respectively; the thrust in BD is 34 pounds' weight.

9. MN is vertically downwards, and the rod is inclined at an angle of $55\frac{1}{2}°$ to the vertical; the tensions of HM, MK, KN, NH are 52, 56, 52, 16 pounds' weight respectively; the thrust in the rod is 60 pounds' weight.

EXAMPLES V.

1. 1·414 P, acting at an angle of 45° with the vertical.

2. 16, 12 pounds' weight.

3. W at each peg, in a direction making 30° with the horizontal.

4. At C, 10 pounds' weight at an angle of 60° with the vertical. At B, 17·32 pounds' weight in a horizontal direction.

5. 44·72 pounds' weight, acting at an angle of $26\frac{1}{2}°$ with the vertical.

6. 39·22 pounds' weight, acting at an angle of $11\frac{1}{3}°$ with the vertical.

7. At B, 11·10 pounds' weight at an angle of $56\frac{1}{3}°$ with the vertical. At C, 16·64 pounds' weight at an angle of $33\frac{2}{3}°$ with the vertical.

8. 41·23 pounds' weight each, at an angle of 14° with the horizontal.

9. 3 pounds' weight; 13·93 pounds' weight.

10. 1·56 pounds' weight.

11. 3·38 ounces' weight.

12. C rests 3 inches below, and 4 inches to the right of, A.
    10 pounds' weight.

13. The ring rests at a point distant 7 inches from A and 25 inches from B, the angle at A being a right angle.
    10 pounds' weight.

14. At B and C, 6 pounds' weight each in a horizontal direction.
    At A, 8 pounds' weight in a vertical direction.
    Tension = 5 pounds' weight.

15. Tension = 10 pounds' weight.
    At B and C, 16 pounds' weight each in a horizontal direction.
    At A, 12 pounds' weight in a vertical direction.

16. Tension = 10 pounds' weight.
    At A, 8·94 pounds' weight at an angle of $63\frac{1}{2}°$ with the vertical.
    At B, 17·89 pounds' weight at an angle of $26\frac{1}{2}°$ with the vertical.
    At C, 16 pounds' weight in a horizontal direction.

17. In the direction CO, where C is in AB at a distance of 3 inches from A.
    Tension = 8·94 pounds' weight.
    At A, 12·65 pounds' weight at an angle of 45° with AB.
    At B, 16·97 pounds' weight at an angle of $18\frac{1}{2}°$ with BA.

18. 95, 65 pounds' weight.

19. 1·75 inches ; 40 pounds' weight.

20. 6 pounds' weight, at an angle of $41\frac{1}{2}°$ with the vertical.

21. When the rod is vertical, the tension of each string is 8 pounds' weight.
    When the rod is horizontal, the tensions of ACB and ADB are 10 and 17 pounds' weight respectively.

EXAMPLES VI.

1. 5, 4 pounds' weight.
2. 17·32, 20 ounces' weight.
3. 36, 20 ounces' weight.
4. 1·15 pounds' weight each.

5. 12 pounds' weight.
6. ·866 W, at an angle of 30° with the vertical.
7. 7·66, 6·43 pounds' weight.
8. (a) 1·82 pounds' weight.
   (b) 1·71 pounds' weight.
9. 6 feet 8 inches.
10. 10 pounds' weight; 31·62 pounds' weight.
11. 40 pounds' weight.
12. 12 ounces' weight.
13. 9·66, 6·73 pounds' weight.
14. 60°, 17·32 pounds' weight. 3·46 inches.
15. 7, 15 ounces' weight.
16. 5·2 ounces' weight.
17. 1 foot, 7·66 pounds' weight.

EXAMPLES VII.

1. (i.) 4·66, 11·03 ounces' weight ;
   (ii.) 4·29, 9·81 ounces' weight ;
   (iii.) 9·41 ounces' weight, at an angle of $23\frac{1}{2}$° with the vertical ;
   (iv.) 4·23 ounces' weight, in an upward direction inclined at an angle of 25° to the horizontal.

2. (i.) 56·7 ounces' weight ;
   (ii.) 5·77 ounces' weight ;
   (iii.) 9·85 ounces' weight, at an angle of 10° with the vertical;
   (iv.) 5 ounces' weight, at an angle of 60° with the vertical.

3. (i.) 7·09 ounces' weight;
   (ii.) 1·92 ounces' weight;
   (iii.) 6·43 ounces' weight, at an angle of 50° with the vertical;
   (iv.) 1·74 ounces' weight, at an angle of 80° with the vertical.

4. 22 ounces' weight ; 1·31 ounces' weight.

5. ·75 ; 3·6 pounds' weight. 2·88 pounds' weight, at an angle of $73\frac{3}{4}$° with the horizontal.

6. 1·15 pounds' weight.

7. 10° and 50° from the highest and lowest points of the hoop.

EXAMPLES VIII.

1. (i.) 12 P, in the same direction as each of the given forces, through a point C in AB, such that $AC =$ ·58 of AB.

   (ii.) 2 P, in the direction of the 7 P, through a point C in AB produced, such that $AC = 3\cdot5$ of AB.

2. 5 pounds' weight, in the direction of the second force, at a distance of 16 inches beyond that force.

3. 34·83 pounds' weight along the line XY, where X is a point in DA, such that $DX =$ ·45 of DA, and Y is a point in CB, such that $CY =$ ·56 of CB.

EXAMPLES IX.

1. 6·93 pounds' weight each, at an angle of 30° with the vertical.

2. 22·5 pounds' weight; 21·93 pounds' weight, at an angle of $24\frac{1}{4}$° with the vertical.

3. 90 pounds' weight.

4. 144 pounds' weight.

5. 45·5 pounds' weight.

6. 17·32 pounds' weight; 17·32 pounds' weight, in a direction perpendicular to the rod.

7. 10 pounds' weight; 17·32 pounds' weight, at an angle of 30° with the vertical.

8. Reaction at $B = \frac{4}{5} W$ in a horizontal direction.
   Reaction at the hinge $= \frac{5}{3} W$ at an angle of 53° with the vertical.

9. 125 pounds' weight; 103·1 pounds' weight, at an angle of 14° with the horizontal. ·25.

10. 13·29 inches, 59·64 ounces' weight.

11. 369·5 pounds' weight, at an angle of $15\frac{2}{3}$° with the horizontal; 260·9 pounds' weight.

12. 1·42 feet from A; 11·66, 17·73 ounces' weight.

13. 2 pounds' weight; 4·47 pounds' weight, at an angle of $26\frac{1}{2}$° with the vertical.

14. 5·46 pounds.

15. 1·86 pounds' weight.

16. ·5 W, ·866 W.

17. W each.

18. 6, 10 pounds' weight.

19. 65 pounds' weight.

20. 26, 28 pounds' weight.

21. 4 inches, 25 pounds' weight.

22. The middle point; $\frac{2}{3}W$, $\frac{4}{8}W$.

23. *Either* from a point dividing AC in the ratio 1 : 3, *or* from a point dividing BC in the ratio 5 : 3; 8·66, 5 pounds' weight.

24. 13 pounds; 19·2 pounds' weight, at an angle of $38\frac{2}{3}°$ with the vertical.

25. 30, 26·46 pounds' weight.

26. Horizontal.

27. Pressure at $D=1$ pound weight, at an angle of 30° with the horizontal.
     Action at $O=1·73$ pounds' weight, at an angle of 60° with the horizontal.

28. Perpendicular to BC, 9 inches from B.

29. In direction OP, where P is at distances of 18 and 24 inches respectively from A and B.

30. 15·6 pounds' weight; 9·92 pounds' weight, at an angle of $52\frac{3}{4}°$ measured downwards from the horizontal.

31. ·5 W, ·577 W.

32. Between 3 and 8 inches from A.

33. 5 inches from A.

34. 1·895 pounds' weight.

35. 96·43, 53·57 pounds' weight.

36. 100, 40 pounds. The first man would support 12 pounds more, and the other 12 pounds less.

37. 9·49 pounds' weight.

38. 16 pounds' weight; 40 pounds' weight.

EXAMPLES X.

1. 1·73 pounds' weight, in a direction perpendicular to the first side, through a point which divides that side externally in the ratio 1 : 3.

2. 96 pounds' weight, in a direction perpendicular to BC, through the middle point of BC.

3. 3·6 pounds' weight along FG, where G is in BC produced such that $CG = \frac{4}{5}.BC$, and F is in AD produced such that $DF = AD$.

4. 105 pounds' weight along LB, where L is in DA produced 9 inches from A.

5. 7 pounds' weight, at an angle of $81\frac{3}{4}°$ with AB, through O in FC produced, where CO = side of hexagon.

EXAMPLES XI.

1. Tension of string = 9 pounds' weight.
   Pressure of step = 15 pounds' weight.
   Pressure of ground = 13 pounds' weight.

2. 8·66, 15, 17·32 pounds' weight; ·58. '

3. − ·5 P, ·71 P, ·5 P.

4. 50, 48·4, 187·5 pounds' weight.

5. 49·24.

6. 5, 12·32, 20 pounds' weight.

7. 15, 12·69, 12·69 pounds' weight.

8. P, 2 P, P.

9. 10, 17·32, 34·64 pounds' weight.

10. 28·87 pounds' weight.
    Tension of AH = 57·74 pounds' weight.
    Tension of BK = 50 pounds' weight.

11. 41·6, 79·2, 36·0 pounds' weight.

12. 25 pounds' weight.

13. 8, 37·86, 8·28 pounds' weight.

14. 10 pounds.
   Tension of BD = 10 pounds' weight.
   Tension of AC = 17·32 pounds' weight.
15. 33, 60, 56 pounds' weight.

EXAMPLES XII.

1. 33·79, 33·21 pounds' weight.
2. $\frac{1}{8}W$.
3. 63 pounds.
4. 5 feet.
5. 3 inches.
6. 6 pounds' weight, 7 inches from A.
7. 2 pounds' weight, 8 inches from A.
8. 20 inches, 12 pounds' weight.
9. 7 pounds' weight, 9·14 inches from A.

EXAMPLES XIII.

1. 45, 330, 45 pounds' weight.
2. 85·33, 270, 85·33 pounds' weight.
3. 25·5 pounds' weight; 154·2 pounds' weight, at an angle of $80\frac{1}{2}°$ with the ground.
4. 70 pounds' weight; 90·4 pounds' weight, at an angle of $27\frac{3}{4}°$ with the vertical.
5. 9 pounds' weight; 15 pounds' weight, at an angle of 37° with the vertical.

EXAMPLES XIV.

1. 20 pounds' weight each, 15 pounds' weight, 25 pounds' weight each.
2. The rod is inclined at an angle of $81\frac{3}{4}°$ to the vertical, H being the highest point; the tensions of LH and LK are 24 and 40 pounds' weight respectively; the thrust in the rod is 15 pounds' weight.

3. 52 pounds' weight. The tensions of the rods BC, CA, AB are 25, 39, 33 pounds' weight respectively; the action at A is 60 pounds' weight, in a direction perpendicular to BC.

4. Tension of string = 115·5 pounds' weight.

, In AB, a thrust of 57·7 pounds' weight.

In BC, a tension of 115·5 pounds' weight.

In CA, a thrust of 115·5 pounds' weight.

Action at A = 152·8 pounds' weight, at an angle of 41° with AB.

5. The force applied at A = 108 pounds' weight.

Reaction at B = 60 pounds' weight vertically upwards.

In BC and CA, tensions of 75 and 117 pounds' weight respectively.

In AB, a thrust of 45 pounds' weight.

6. 28 pounds' weight each.

In AB, BC, CD, DA, tensions of 30, 26, 26, 30 pounds' weight respectively.

7. 80 pounds' weight each.

8. The forces in the lines bc, da, and the tensions of the rods oa, ob, oc, od are 80, 60, 36, 48, 64, 48 pounds' weight respectively.

9. Tension of string = 200 pounds' weight.

In AB, BC, CD, DA, tensions of 130, 90, 130, 90 pounds' weight respectively.

10. In AB and EF, tensions of 51 pounds' weight each.

In BC and DE, tensions of 30 pounds' weight each.

In CD, tension of 18 pounds' weight.

In BE, tension of 27 pounds' weight.

11. In AC, CD, DB, tensions of 20, 12, 20 pounds' weight respectively.

12. In AB and BC, tensions of 25 and 50 pounds' weight respectively.

In AD and DC, thrusts of 25 and 50 pounds' weight respectively. The reactions at B and D are 49·24 pounds' weight each.

13. The actions at B and D are 48 and 161·3 pounds' weight respectively.

In AD and DC, tensions of 250 and 300 pounds' weight respectively.

In AB and BC, thrusts of 240 and 288 pounds' weight respectively.

14. In AB, BC, CD, DE, tensions of 56, 65, 65, 56 pounds' weight respectively.

In BF and FD, thrusts of 33 pounds' weight each.

EXAMPLES XV.

1. In AB, BC, CD, tensions of $1·155\,W$, $·577\,W$, $1·155\,W$ respectively, where W is the weight of each mass.

2. 8·66 pounds' weight.
   In AB, a thrust of 3 pounds' weight.
   In BC and CD, tensions of 5 pounds' weight each.

3. 63 pounds' weight.
   In BC and CD, tensions of 52 and 25 pounds' weight respectively.
   In AB, a thrust of 20 pounds' weight.

4. 69·29 pounds' weight.
   In BC and CD, tensions of 34·6 and 60 pounds' weight respectively.
   In AB, a thrust of 17·3 pounds' weight.

5. 33·3 pounds.
   In AB, BC, CD, tensions of 115·5, 57·7, 66·7 pounds' weight respectively.

6. (i.) $59\frac{1}{4}°$, $90°$, $59\frac{1}{4}°$ to the vertical;
   (ii.) 11·64, 10, 11·64 feet;
   (iii.) 48·85, 41·95, 48·85 pounds' weight.

7. (i.) $25°$;
   (ii.) 6627 pounds' weight;
   (iii.) 2813 pounds' weight;
   (iv.) 4·56 feet.

EXAMPLES XVI.

1. In AC, a tension of 112 pounds' weight.
   In AB and AD tensions, in BC and CD thrusts, each of 64 pounds' weight.

2. In AC, a tension of 52 pounds' weight.
   In AB and AD, tensions of 80 pounds' weight each.
   In BC and CD, thrusts of 40 pounds' weight each.

3. In DC, a thrust of 272 pounds' weight.
   In BC, a tension of 308 pounds' weight.
   In DB, a thrust of 231 pounds' weight.
   In AB, a tension of 385 pounds' weight.

4. AB is horizontal.
   In AC, AB, DA, tensions of 8·75, 2·04, 26·04 pounds' weight respectively.
   In BC and CD, thrusts of 7·29 pounds' weight each.

5. 192 pounds.
   In AC, a tension of 150 pounds' weight.
   In AB a tension and in BC a thrust, each of 90 pounds' weight.
   In AD a tension and in DC a thrust, each of 120 pounds' weight.

6. 50 pounds' weight.
   In KH, KM, KL, tensions of 30, 50, 40 pounds' weight respectively.
   In MH and ML, thrusts of 40 and 30 pounds' weight respectively.

7. 32 pounds' weight.
   In BA, BD, BC, tensions of 50, 56, 34 pounds' weight respectively.
   In DA and DC, thrusts of 50 and 34 pounds' weight respectively.

8. The forces applied at A and C are 120 pounds' weight each.
   In BA, BD, BC, tensions of 125, 42, 125 pounds' weight respectively.
   In DA and DC, thrusts of 35 pounds' weight each.
   D.S.   Y

9. In AB, BC, CD, CA, tensions of 11·55, 5·77, 28·87, 10 pounds' weight respectively.

10. In AB, BC, CD, CA, thrusts of 11·55, 5·77, 28·87, 10 pounds' weight respectively.
In AD, a tension of 14·43 pounds' weight.
The reactions at A and D are 15 and 25 pounds' weight respectively.

11. In the lower rods, tensions of 120 pounds' weight each.
In the upper rods, thrusts of 130 pounds' weight each.
In the upright, a tension of 100 pounds' weight.

12. In the lower rods, tensions of 125 pounds' weight each.
In the upper rods, thrusts of 150 pounds' weight each. ·
In the upright, a tension of 180 pounds' weight.

13. In the 28-foot rod, a thrust of 14,000 pounds' weight.
In the 20-foot rod, a tension of 10,000 pounds' weight.
In the upright, a thrust of 15,780 pounds' weight.
In the 13-foot rod, a tension of 22,520 pounds' weight.
The action at A is 28,150 pounds' weight, at an angle of 18° with the vertical.

EXAMPLES XVII.

1. In BC, a thrust of 117 pounds' weight.
In OB and OC, ties of 97·5 pounds' weight each.
In OA and OD, ties of 62·5 pounds' weight each.
In BA and CD, thrusts of 97·5 pounds' weight each.

2. In the two lower rods, tensions of 288·7 pounds' weight each.
In the three upper rods, thrusts of 577·4 pounds' weight each.
In the two internal rods, tensions of 577·4 pounds' weight each.

3. In the two lower rods, tensions of 150 pounds' weight each.
In the two side rods, thrusts of 250 pounds' weight each.
In the top middle rod, a thrust of 300 pounds' weight.
In the two internal rods, tensions of 250 pounds' weight each.

5. Tension of 808·9 pounds' weight, thrust of 622·2 pounds' weight, tension of 311·1 pounds' weight.

7. In the lower horizontal rods, tensions of 466·7, 1400, 1400, 466·7 pounds' weight.
In the upper horizontal rods, thrusts of 933·3, 1866·7, 933·3 pounds' weight.
The other rods, taken in order, are:—strut, tie, strut, tie, tie, strut, tie, strut.

9. In the top rod, a thrust of 4480 pounds' weight.
In the bottom rod, a tension of 5973 pounds' weight.
In the internal rod, a thrust of 1867 pounds' weight.

12. In the horizontal rod, a tension of 112 pounds' weight.
In the top rod, a thrust of 200 pounds' weight.
In the internal rod, a tension of 80 pounds' weight.

EXAMPLES XVIII.

1. 26, 22·5, 22·5, 26 ounces' weight.

2. In HK, HF, HN, tensions of 26, 36, 26 ounces' weight respectively.
The actions at L and M are each 26 ounces' weight, in directions parallel to HN, HK respectively.

3. 2·5, 4·2, 4·2, 2·5 pounds' weight.

4. Thrust in the cross-rod = 57 pounds' weight.
Tension of each string = 25 pounds' weight.
Action at the hinge   = 37 pounds' weight, in a horizontal direction.

5. Thrust in the cross-rod = 3·46 W.
The actions at the hinges are:—
At A and F, 2·75 W each, at an angle of $24\frac{3}{4}°$ with the vertical.
At B and E, 1·89 W each, at an angle of $37\frac{1}{2}°$ with the vertical.
At C and D, 2·36 W each, at an angle of $77\frac{3}{4}°$ with the vertical.

6. Tension of string = 1·15 W.
Action at H = ·5 W, in a vertical direction.
Action at L = ·76 W, at an angle of 49° with the vertical.

7. 1·5 W; ·29 W, 1·73 W, 2·31 W.

8. Tension of string = 4 W.
   Action at K = ·87 W, in a horizontal direction.
   Action at L = 1·32 W, at an angle of 41° with the vertical.
   Action at M = 2·18 W, at an angle of 23½° with the vertical.

9. Tension of string = 2·67 W.
   Action at K = ·87 W, in a horizontal direction.
   Action at L = 1·32 W, at an angle of 41° with the vertical.
   Action at M = 1·09 W, at an angle of 52½° with the vertical.

10. 28 pounds' weight each.

11. Tension of FH = 21 pounds' weight.
    Tension of GK = 35 pounds' weight.

EXAMPLES XIX.

1. 4·9 pounds' weight.
2. 30, 26·46 pounds' weight.
3. 126·59, 136·85 pounds' weight.
5. 115·5 pounds' weight.
6. W.

EXAMPLES XX.

1. ·164.
2. ·5 W, 30° with the horizontal.   60°.
3. 50 feet up the ladder.
4. 16¾°.
5. To the top of the ladder.
6. 45°.
7. It slips.

GLASGOW : PRINTED AT THE UNIVERSITY PRESS BY ROBERT MACLEHOSE AND CO.

www.ingramcontent.com/pod-product-compliance
Lightning Source LLC
Chambersburg PA
CBHW021404210326
41599CB00011B/1005